Springer Texts in Statistics

Advisors:
Stephen Fienberg   Ingram Olkin

Springer Texts in Statistics

Stephen Kokoska   Christopher Nevison

Statistical Tables
and Formulae

Springer-Verlag
New York Berlin Heidelberg
London Paris Tokyo

Stephen Kokoska
Department of Mathematics
Colgate University
Hamilton, NY 13346-1398
USA

Christopher Nevison
Department of Computer Science
Colgate University
Hamilton, NY 13346-1398
USA

Editorial Board

Stephen Fienberg
Department of Statistics
Carnegie-Mellon University
Pittsburgh, PA 15213
USA

Ingram Olkin
Department of Statistics
Stanford University
Stanford, CA 94305
USA

Mathematics Subject Classification (1980): 62Q05

Library of Congress Cataloging-in-Publication Data
Kokoska, Stephen.
  (Springer texts in statistics)
  1. Mathematical statistics--Tables. 2. Probabilities
--Tables. I. Title. II. Series.
QA276.25.K65   1988   519.5'0212   88-24955

Printed on acid-free paper

Camera-ready copy provided by the authors.
Printed and bound by Edwards Brothers, Inc., Ann Arbor, Michigan.
Printed in the United States of America.

9 8 7 6 5 4 3 2 1

ISBN 0-387-96873-3 Springer-Verlag New York Berlin Heidelberg
ISBN 3-540-96873-3 Springer-Verlag Berlin Heidelberg New York

Contents

Table 1. Discrete Distributions

Probability Mass Function, $p(x)$; Mean, μ; Variance, σ^2; Coefficient of Skewness, β_1;
Coefficient of Kurtosis, β_2; Moment-generating Function, $M(t)$; Characteristic Function, $\phi(t)$;
Probability-generating Function, $P(t)$.

Bernoulli Distribution

$$p(x) = p^x q^{x-1} \qquad x = 0, 1 \qquad 0 \le p \le 1 \qquad q = 1 - p$$

$$\mu = p \qquad \sigma^2 = pq \qquad \beta_1 = \frac{1 - 2p}{\sqrt{pq}} \qquad \beta_2 = 3 + \frac{1 - 6pq}{pq}$$

$$M(t) = q + pe^t \qquad \phi(t) = q + pe^{it} \qquad P(t) = q + pt$$

Beta Binomial Distribution

$$p(x) = \frac{1}{n+1} \frac{B(a + x, b + n - x)}{B(x + 1, n - x + 1)B(a, b)} \qquad x = 0, 1, 2, \ldots, n \qquad a > 0 \qquad b > 0$$

$$\mu = \frac{na}{a + b} \qquad \sigma^2 = \frac{nab(a + b + n)}{(a + b)^2(a + b + 1)} \qquad B(a, b) \text{ is the Beta function.}$$

Beta Pascal Distribution

$$p(x) = \frac{\Gamma(x)\Gamma(\nu)\Gamma(\rho + \nu)\Gamma(\nu + x - (\rho + r))}{\Gamma(r)\Gamma(x - r + 1)\Gamma(\rho)\Gamma(\nu - \rho)\Gamma(\nu + x)} \qquad x = r, r + 1, \ldots \qquad \nu > \rho > 0$$

$$\mu = r\frac{\nu - 1}{\rho - 1}, \ \rho > 1 \qquad \sigma^2 = r(r + \rho - 1)\frac{(\nu - 1)(\nu - \rho)}{(\rho - 1)^2(\rho - 2)}, \ \rho > 2$$

Binomial Distribution

$$p(x) = \binom{n}{x} p^x q^{n-x} \qquad x = 0, 1, 2, \ldots, n \qquad 0 \le p \le 1 \qquad q = 1 - p$$

$$\mu = np \qquad \sigma^2 = npq \qquad \beta_1 = \frac{1 - 2p}{\sqrt{npq}} \qquad \beta_2 = 3 + \frac{1 - 6pq}{npq}$$

$$M(t) = (q + pe^t)^n \qquad \phi(t) = (q + pe^{it})^n \qquad P(t) = (q + pt)^n$$

Discrete Weibull Distribution

$$p(x) = (1 - p)^{x^\beta} - (1 - p)^{(x+1)^\beta} \qquad x = 0, 1, \ldots \qquad 0 \le p \le 1 \qquad \beta > 0$$

Geometric Distribution

$$p(x) = pq^{1-x} \qquad x = 0, 1, 2 \ldots \qquad 0 \le p \le 1 \qquad q = 1 - p$$

$$\mu = \frac{1}{p} \qquad \sigma^2 = \frac{q}{p^2} \qquad \beta_1 = \frac{2 - p}{\sqrt{q}} \qquad \beta_2 = \frac{p^2 + 6q}{q}$$

$$M(t) = \frac{p}{1 - qe^t} \qquad \phi(t) = \frac{p}{1 - qe^{it}} \qquad P(t) = \frac{p}{1 - qt}$$

Table 1. Discrete Distributions (Continued)

Hypergeometric Distribution

$$p(x) = \frac{\binom{M}{x}\binom{N-M}{n-x}}{\binom{N}{n}} \qquad x = 0, 1, 2, \ldots, n \qquad x \le M \qquad n - x \le N - M$$

$$n, M, N \in \mathbf{N} \qquad 1 \le n \le N \qquad 1 \le M \le N \qquad N = 1, 2, \ldots$$

$$\mu = n\frac{M}{N} \qquad \sigma^2 = \left(\frac{N-n}{N-1}\right) n\frac{M}{N}\left(1 - \frac{M}{N}\right) \qquad \beta_1 = \frac{(N-2M)(N-2n)\sqrt{N-1}}{(N-2)\sqrt{nM(N-M)(N-n)}}$$

$$\beta_2 = \frac{N^2(N-1)}{(N-2)(N-3)nM(N-M)(N-n)} \cdot$$

$$\left\{ N(N+1) - 6n(N-n) + 3\tfrac{M}{N^2}(N-M)[N^2(n-2) - Nn^2 + 6n(N-n)] \right\}$$

$$M(t) = \frac{(N-M)!(N-n)!}{N!}F(\cdot, e^t) \qquad \phi(t) = \frac{(N-M)!(N-n)!}{N!}F(\cdot, e^{it}) \qquad P(t) = \left(\frac{N-M}{N}\right)^n F(\cdot, t)$$

$F(\alpha, \beta, \gamma, x)$ is the hypergeometric function. $\qquad \alpha = -n; \quad \beta = -M; \quad \gamma = N - M - n + 1$

Negative Binomial Distribution

$$p(x) = \binom{x+r-1}{r-1} p^r q^x \qquad x = 0, 1, 2, \ldots \qquad r = 1, 2, \ldots \qquad 0 \le p \le 1 \qquad q = 1 - p$$

$$\mu = \frac{rq}{p} \qquad \sigma^2 = \frac{rq}{p^2} \qquad \beta_1 = \frac{2-p}{\sqrt{rq}} \qquad \beta_2 = 3 + \frac{p^2 + 6q}{rq}$$

$$M(t) = \left(\frac{p}{1-qe^t}\right)^r \qquad \phi(t) = \left(\frac{p}{1-qe^{it}}\right)^r \qquad P(t) = \left(\frac{p}{1-qt}\right)^r$$

Poisson Distribution

$$p(x) = \frac{e^{-\mu}\mu^x}{x!} \qquad x = 0, 1, 2, \ldots \qquad \mu > 0$$

$$\mu = \mu \qquad \sigma^2 = \mu \qquad \beta_1 = \frac{1}{\sqrt{\mu}} \qquad \beta_2 = 3 + \frac{1}{\mu}$$

$$M(t) = \exp[\mu(e^t - 1)] \qquad \phi(t) = \exp[\mu(e^{it} - 1)] \qquad P(t) = \exp[\mu(t-1)]$$

Rectangular (Discrete Uniform) Distribution

$$p(x) = 1/n \qquad x = 1, 2, \ldots, n \qquad n \in \mathbf{N}$$

$$\mu = \frac{n+1}{2} \qquad \sigma^2 = \frac{n^2-1}{12} \qquad \beta_1 = 0 \qquad \beta_2 = \frac{3}{5}\left(3 - \frac{4}{n^2-1}\right)$$

$$M(t) = \frac{e^t(1-e^{nt})}{n(1-e^t)} \qquad \phi(t) = \frac{e^{it}(1-e^{nit})}{n(1-e^{it})} \qquad P(t) = \frac{t(1-t^n)}{n(1-t)}$$

Table 2. Continuous Distributions

Probability Density Function, $f(x)$; Mean, μ; Variance, σ^2; Coefficient of Skewness, β_1; Coefficient of Kurtosis, β_2; Moment-generating Function, $M(t)$; Characteristic Function, $\phi(t)$.

Arcsin Distribution

$$f(x) = \frac{1}{\pi\sqrt{x(1-x)}} \qquad 0 < x < 1$$

$$\mu = \frac{1}{2} \qquad \sigma^2 = \frac{1}{8} \qquad \beta_1 = 0 \qquad \beta_2 = \frac{3}{2}$$

Beta Distribution

$$f(x) = \frac{\Gamma(\alpha+\beta)}{\Gamma(\alpha)\Gamma(\beta)} x^{\alpha-1}(1-x)^{\beta-1} \qquad 0 < x < 1 \qquad \alpha, \ \beta > 0$$

$$\mu = \frac{\alpha}{\alpha+\beta} \qquad \sigma^2 = \frac{\alpha\beta}{(\alpha+\beta)^2(\alpha+\beta+1)} \qquad \beta_1 = \frac{2(\beta-\alpha)\sqrt{\alpha+\beta+1}}{\sqrt{\alpha\beta}(\alpha+\beta+2)}$$

$$\beta_2 = \frac{3(\alpha+\beta+1)[2(\alpha+\beta)^2+\alpha\beta(\alpha+\beta-6)]}{\alpha\beta(\alpha+\beta+2)(\alpha+\beta+3)}$$

Cauchy Distribution

$$f(x) = \frac{1}{b\pi\left(1+\left(\frac{x-a}{b}\right)^2\right)} \qquad -\infty < x < \infty \qquad -\infty < a < \infty \qquad b > 0$$

μ, σ^2, β_1, β_2, $M(t)$ do not exist. $\qquad \phi(t) = \exp[ait - b\,|\,t\,|]$

Chi Distribution

$$f(x) = \frac{x^{n-1}e^{-x^2/2}}{2^{(n/2)-1}\Gamma(n/2)} \qquad x \geq 0 \qquad n \in \mathbf{N}$$

$$\mu = \frac{\Gamma\left(\frac{n+1}{2}\right)}{\Gamma\left(\frac{n}{2}\right)} \qquad \sigma^2 = \frac{\Gamma\left(\frac{n+2}{2}\right)}{\Gamma\left(\frac{n}{2}\right)} - \left[\frac{\Gamma\left(\frac{n+1}{2}\right)}{\Gamma\left(\frac{n}{2}\right)}\right]^2$$

Chi-Square Distribution

$$f(x) = \frac{e^{-x/2}x^{(\nu/2)-1}}{2^{\nu/2}\Gamma(\nu/2)} \qquad x \geq 0 \qquad \nu \in \mathbf{N}$$

$$\mu = \nu \qquad \sigma^2 = 2\nu \qquad \beta_1 = 2\sqrt{2/\nu} \qquad \beta_2 = 3 + \frac{12}{\nu} \qquad M(t) = (1-2t)^{-\nu/2}, \ t < \frac{1}{2} \qquad \phi(t) = (1-2it)^{-\nu/2}$$

Erlang Distribution

$$f(x) = \frac{1}{\beta^n(n-1)!}x^{n-1}e^{-x/\beta} \qquad x \geq 0 \qquad \beta > 0 \qquad n \in \mathbf{N}$$

$$\mu = n\beta \qquad \sigma^2 = n\beta^2 \qquad \beta_1 = \frac{2}{\sqrt{n}} \qquad \beta_2 = 3 + \frac{6}{n} \qquad M(t) = (1-\beta t)^{-n} \qquad \phi(t) = (1-\beta it)^{-n}$$

Table 2. Continuous Distributions (Continued)

Exponential Distribution

$$f(x) = \lambda e^{-\lambda x} \qquad x \geq 0 \qquad \lambda > 0$$

$$\mu = \frac{1}{\lambda} \qquad \sigma^2 = \frac{1}{\lambda^2} \qquad \beta_1 = 2 \qquad \beta_2 = 9 \qquad M(t) = \frac{\lambda}{\lambda - t} \qquad \phi(t) = \frac{\lambda}{\lambda - it}$$

Extreme-Value Distribution

$$f(x) = \exp\left[-e^{-(x-\alpha)/\beta}\right] \qquad -\infty < x < \infty \qquad -\infty < \alpha < \infty \qquad \beta > 0$$

$$\mu = \alpha + \gamma\beta, \ \gamma \doteq .5772\ldots \text{ is Euler's constant} \qquad \sigma^2 = \frac{\pi^2 \beta^2}{6} \qquad \beta_1 = 1.29857 \qquad \beta_2 = 5.4$$

$$M(t) = e^{\alpha t}\Gamma(1 - \beta t), \ t < \frac{1}{\beta} \qquad \phi(t) = e^{\alpha it}\Gamma(1 - \beta it)$$

F Distribution

$$f(x) = \frac{\Gamma[(\nu_1 + \nu_2)/2]\nu_1^{\nu_1/2}\nu_2^{\nu_2/2}}{\Gamma(\nu_1/2)\Gamma(\nu_2/2)}x^{(\nu_1/2)-1}(\nu_2 + \nu_1 x)^{-(\nu_1 + \nu_2)/2} \qquad x > 0 \qquad \nu_1, \nu_2 \in \mathbf{N}$$

$$\mu = \frac{\nu_2}{\nu_2 - 2}, \ \nu_2 \geq 3 \qquad \sigma^2 = \frac{2\nu_2^2(\nu_1 + \nu_2 - 2)}{\nu_1(\nu_2 - 2)^2(\nu_2 - 4)}, \ \nu_2 \geq 5$$

$$\beta_1 = \frac{(2\nu_1 + \nu_2 - 2)\sqrt{8(\nu_2 - 4)}}{\sqrt{\nu_1}(\nu_2 - 6)\sqrt{\nu_1 + \nu_2 - 2}}, \ \nu_2 \geq 7$$

$$\beta_2 = 3 + \frac{12[(\nu_2 - 2)^2(\nu_2 - 4) + \nu_1(\nu_1 + \nu_2 - 2)(5\nu_2 - 22)]}{\nu_1(\nu_2 - 6)(\nu_2 - 8)(\nu_1 + \nu_2 - 2)}, \ \nu_2 \geq 9$$

$$M(t) \text{ does not exist.} \qquad \phi(\frac{\nu_1}{\nu_2}t) = \frac{G(\nu_1, \nu_2, t)}{B(\nu_1/2, \nu_2/2)}$$

$B(a, b)$ is the Beta function. G is defined by

$$(m + n - 2)G(m, n, t) = (m - 2)G(m - 2, n, t) + 2itG(m, n - 2, t), \ m, n > 2$$

$$mG(m, n, t) = (n - 2)G(m + 2, n - 2, t) - 2itG(m + 2, n - 4, t), \ n > 4$$

$$nG(2, n, t) = 2 + 2itG(2, n - 2, t), \ n > 2$$

Gamma Distribution

$$f(x) = \frac{1}{\beta^\alpha \Gamma(\alpha)}x^{\alpha - 1}e^{-x/\beta} \qquad x \geq 0 \qquad \alpha, \beta > 0$$

$$\mu = \alpha\beta \qquad \sigma^2 = \alpha\beta^2 \qquad \beta_1 = \frac{2}{\sqrt{\alpha}} \qquad \beta_2 = 3\left(1 + \frac{2}{\alpha}\right) \qquad M(t) = (1 - \beta t)^{-\alpha} \qquad \phi(t) = (1 - \beta it)^{-\alpha}$$

Table 2. Continuous Distributions (Continued)

Half-Normal Distribution

$$f(x) = \frac{2\theta}{\pi} \exp[-(\theta^2 x^2/\pi)] \qquad x \geq 0 \qquad \theta > 0$$

$$\mu = \frac{1}{\theta} \qquad \sigma^2 = \left(\frac{\pi - 2}{2}\right)\frac{1}{\theta^2} \qquad \beta_1 = \frac{4 - \pi}{\theta^3} \qquad \beta_2 = \frac{3\pi^2 - 4\pi - 12}{4\theta^4}$$

LaPlace (Double Exponential) Distribution

$$f(x) = \frac{1}{2\beta} \exp\left[-\frac{|x - \alpha|}{\beta}\right] \qquad -\infty < x < \infty \qquad -\infty < \alpha < \infty \qquad \beta > 0$$

$$\mu = \alpha \qquad \sigma^2 = 2\beta^2 \qquad \beta_1 = 0 \qquad \beta_2 = 6 \qquad M(t) = \frac{e^{\alpha t}}{1 - \beta^2 t^2} \qquad \phi(t) = \frac{e^{\alpha it}}{1 + \beta^2 t^2}$$

Logistic Distribution

$$f(x) = \frac{\exp[(x - \alpha)/\beta]}{\beta(1 + \exp[(x - \alpha)/\beta])^2} \qquad -\infty < x < \infty \qquad -\infty < \alpha < \infty \qquad -\infty < \beta < \infty$$

$$\mu = \alpha \qquad \sigma^2 = \frac{\beta^2 \pi^2}{3} \qquad \beta_1 = 0 \qquad \beta_2 = 4.2 \qquad M(t) = e^{\alpha t}\pi\beta t \csc(\pi\beta t) \qquad \phi(t) = e^{\alpha it}\pi\beta it \csc(\pi\beta it)$$

Lognormal Distribution

$$f(x) = \frac{1}{\sqrt{2\pi}\sigma x} \exp\left[-\frac{1}{2\sigma^2}(\ln x - \mu)^2\right] \qquad x > 0 \qquad -\infty < \mu < \infty \qquad \sigma > 0$$

$$\mu = e^{\mu + \sigma^2/2} \qquad \sigma^2 = e^{2\mu + \sigma^2}(e^{\sigma^2} - 1) \qquad \beta_1 = (e^{\sigma^2} + 2)(e^{\sigma^2} - 1)^{1/2} \qquad \beta_2 = (e^{\sigma^2})^4 + 2(e^{\sigma^2})^3 + 3(e^{\sigma^2})^2 - 3$$

Noncentral Chi-Square Distribution

$$f(x) = \frac{\exp\left[-\frac{1}{2}(x + \lambda)\right]}{2^{\nu/2}} \sum_{j=0}^{\infty} \frac{x^{(\nu/2)+j-1}\lambda^j}{\Gamma\left(\frac{\nu}{2} + j\right)2^{2j}j!} \qquad x > 0 \qquad \lambda > 0 \qquad \nu \in \mathbf{N}$$

$$\mu = \nu + \lambda \qquad \sigma^2 = 2(\nu + 2\lambda) \qquad \beta_1 = \frac{\sqrt{8}(\nu + 3\lambda)}{(\nu + 2\lambda)^{3/2}} \qquad \beta_2 = 3 + \frac{12(\nu + 4\lambda)}{(\nu + 2\lambda)^2}$$

$$M(t) = (1 - 2t)^{-\nu/2} \exp\left[\frac{\lambda t}{1 - 2t}\right] \qquad \phi(t) = (1 - 2it)^{-\nu/2} \exp\left[\frac{\lambda it}{1 - 2it}\right]$$

Noncentral F Distribution

$$f(x) = \sum_{i=0}^{\infty} \frac{\Gamma\left(\frac{2i+\nu_1+\nu_2}{2}\right)\left(\frac{\nu_1}{\nu_2}\right)^{(2i+\nu_1)/2} x^{(2i+\nu_1-2)/2} e^{-\lambda/2}\left(\frac{\lambda}{2}\right)}{\Gamma\left(\frac{\nu_2}{2}\right)\Gamma\left(\frac{2i+\nu_1}{2}\right)\nu_1!\left(1 + \frac{\nu_1}{\nu_2}x\right)^{(2i+\nu_1+\nu_2)/2}} \qquad x > 0 \qquad \nu_1, \nu_2 \in \mathbf{N} \qquad \lambda > 0$$

$$\mu = \frac{(\nu_1 + \lambda)\nu_2}{(\nu_2 - 2)\nu_1}, \ \nu_2 > 2 \qquad \sigma^2 = \frac{(\nu_1 + \lambda)^2 + 2(\nu_1 + \lambda)\nu_2^2}{(\nu_2 - 2)(\nu_2 - 4)\nu_1^2} - \frac{(\nu_1 + \lambda)^2\nu_2^2}{(\nu_2 - 2)^2\nu_1^2}, \ \nu_2 > 4$$

Table 2. Continuous Distributions (Continued)

Noncentral t Distribution

$$f(x) = \frac{\nu^{\nu/2}}{\Gamma\left(\frac{\nu}{2}\right)} \frac{e^{-\delta^2/2}}{\sqrt{\pi}(\nu + x^2)^{(\nu+1)/2}} \sum_{i=0}^{\infty} \Gamma\left(\frac{\nu+i+1}{2}\right) \left(\frac{\delta^i}{i!}\right) \left(\frac{2x^2}{\nu + x^2}\right)^{i/2}$$

$$-\infty < x < \infty \qquad -\infty < \delta < \infty \qquad \nu \in \mathbf{N}$$

$$\mu_r' = c_r \frac{\Gamma\left(\frac{\nu-r}{2}\right)\nu^{r/2}}{2^{r/2}\Gamma\left(\frac{\nu}{2}\right)}, \ \nu > r, \quad c_{2r-1} = \sum_{i=1}^{r} \frac{(2r-1)!\delta^{2r-1}}{(2i-1)!(r-i)!2^{r-i}}, \quad c_{2r} = \sum_{i=0}^{r} \frac{(2r)!\delta^{2i}}{(2i)!(r-i)!2^{r-i}}, \quad r = 1,2,3,\ldots$$

Normal Distribution

$$f(x) = \frac{1}{\sigma\sqrt{2\pi}} \exp\left[-\frac{(x-\mu)^2}{2\sigma^2}\right] \qquad -\infty < x < \infty \qquad -\infty < \mu < \infty \qquad \sigma > 0$$

$$\mu = \mu \qquad \sigma^2 = \sigma^2 \qquad \beta_1 = 0 \qquad \beta_2 = 3 \qquad M(t) = \exp\left[\mu t + \frac{t^2\sigma^2}{2}\right] \qquad \phi(t) = \exp\left[\mu i t - \frac{t^2\sigma^2}{2}\right]$$

Pareto Distribution

$$f(x) = \theta a^\theta / x^{\theta+1} \qquad x \geq a \qquad \theta > 0 \qquad a > 0$$

$$\mu = \frac{\theta a}{\theta - 1}, \ \theta > 1 \qquad \sigma^2 = \frac{\theta a^2}{(\theta-1)^2(\theta-2)}, \ \theta > 2$$

$$\beta_1 = \frac{2(\theta+1)}{(\theta-3)(\theta-1)\sqrt{\theta(\theta-2)}}, \ \theta > 3 \qquad \beta_2 = \frac{3(\theta-2)(3\theta^2+\theta+2)}{\theta(\theta-3)(\theta-4)}, \ \theta > 4$$

$M(t)$ does not exist.

Rayleigh Distribution

$$f(x) = \frac{x}{\sigma^2} \exp\left[-\frac{x^2}{2\sigma^2}\right] \qquad x \geq 0 \qquad \sigma > 0$$

$$\mu = \sigma\sqrt{\pi/2} \qquad \sigma^2 = 2\sigma^2\left(1 - \frac{\pi}{4}\right) \qquad \beta_1 = \frac{\sqrt{\pi}}{4}\frac{(\pi-3)}{(1-\frac{\pi}{4})^{3/2}} \qquad \beta_2 = \frac{2 - \frac{3}{16}\pi^2}{(1-\frac{\pi}{4})^2}$$

t Distribution

$$f(x) = \frac{1}{\sqrt{\pi\nu}} \frac{\Gamma(\frac{\nu+1}{2})}{\Gamma(\frac{\nu}{2})} \left(1 + \frac{x^2}{\nu}\right)^{-(\nu+1)/2} \qquad -\infty < x < \infty \qquad \nu \in \mathbf{N}$$

$$\mu = 0, \ \nu \geq 2 \qquad \sigma^2 = \frac{\nu}{\nu - 2}, \ \nu \geq 3 \qquad \beta_1 = 0, \ \nu \geq 4 \qquad \beta_2 = 3 + \frac{6}{\nu - 4}, \ \nu \geq 5$$

$$M(t) \text{ does not exist.} \qquad \phi(t) = \frac{\sqrt{\pi}\Gamma(\frac{\nu}{2})}{\Gamma(\frac{\nu+1}{2})} \int_{-\infty}^{\infty} \frac{e^{itz\sqrt{\nu}}}{(1+z^2)^{(\nu+1)/2}} \, dz$$

Table 2. Continuous Distributions (Continued)

Triangular Distribution

$$f(x) = \begin{cases} 0 & x \leq a \\ 4(x-a)/(b-a)^2 & a < x \leq (a+b)/2 \\ 4(b-x)/(b-a)^2 & (a+b)/2 < x < b \\ 0 & x \geq b \end{cases} \quad -\infty < a < b < \infty$$

$$\mu = \frac{a+b}{2} \qquad \sigma^2 = \frac{(b-a)^2}{24} \qquad \beta_1 = 0 \qquad \beta_2 = \frac{12}{5}$$

$$M(t) = -\frac{4(e^{at/2} - e^{bt/2})^2}{t^2(b-a)^2} \qquad \phi(t) = \frac{4(e^{ait/2} - e^{bit/2})^2}{t^2(b-a)^2}$$

Uniform Distribution

$$f(x) = \frac{1}{b-a} \qquad a \leq x \leq b \qquad -\infty < a < b < \infty$$

$$\mu = \frac{a+b}{2} \qquad \sigma^2 = \frac{(b-a)^2}{12} \qquad \beta_1 = 0 \qquad \beta_2 = \frac{9}{5}$$

$$M(t) = \frac{e^{bt} - e^{at}}{(b-a)t} \qquad \phi(t) = \frac{e^{bit} - e^{ait}}{(b-a)it}$$

Weibull Distribution

$$f(x) = \frac{\alpha}{\beta^\alpha} x^{\alpha-1} e^{-(x/\beta)^\alpha} \qquad x \geq 0 \qquad \alpha, \beta > 0$$

$$\mu = \beta\Gamma\left(1+\frac{1}{\alpha}\right) \qquad \sigma^2 = \beta^2\left[\Gamma\left(1+\frac{2}{\alpha}\right) - \Gamma^2\left(1+\frac{1}{\alpha}\right)\right]$$

$$\beta_1 = \frac{\Gamma(1+\frac{3}{\alpha}) - 3\Gamma(1+\frac{1}{\alpha})\Gamma(1+\frac{2}{\alpha}) + 2\Gamma^3(1+\frac{1}{\alpha})}{[\Gamma(1+\frac{2}{\alpha}) - \Gamma^2(1+\frac{1}{\alpha})]^{3/2}}$$

$$\beta_2 = \frac{\Gamma(1+\frac{4}{\alpha}) - 4\Gamma(1+\frac{1}{\alpha})\Gamma(1+\frac{3}{\alpha}) + 6\Gamma^2(1+\frac{1}{\alpha})\Gamma(1+\frac{2}{\alpha}) - 3\Gamma^4(1+\frac{1}{\alpha})}{[\Gamma(1+\frac{2}{\alpha}) - \Gamma^2(1+\frac{1}{\alpha})]^2}$$

Key To Table 3.

Table 3 presents some of the relationships among common univariate distributions. The first line in each box is the name of the distribution and the second line lists the distribution's parameters. Parameter restrictions and the values each random variable takes on with positive probability are given in Tables 1 and 2. The random variable X is used to represent each distribution. The three types of relationships represented in the diagram are transformations (independent random variables are assumed) and special cases (both indicated with a solid arrow), and limiting distributions (indicated with a dashed arrow).

Table 3. Relationships Among Distributions

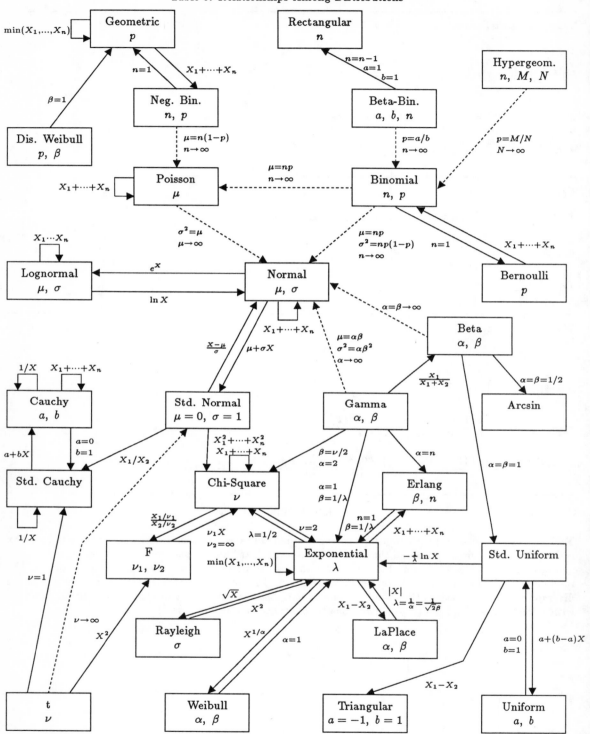

Table 4. Probability and Statistics Formulas

Combinatorial Methods

The Product Rule for Ordered Pairs: If the first element of an ordered pair can be selected in n_1 ways, and for each of these n_1 ways the second element of the pair can be selected in n_2 ways, then the number of possible pairs is $n_1 n_2$.

The Generalized Product Rule for k-tuples: If a set consists of ordered collections of k-tuples and there are n_1 choices for the first element; and for each choice of the first element there are n_2 choices for the second element; ... ; and for each choice of the first $k-1$ elements there are n_k choices for the kth element, then there are $n_1 n_2 \cdots n_k$ possible k-tuples.

Permutations: The number of permutations of n distinct objects taken k at a time is $P_{n,k} = \dfrac{n!}{(n-k)!}$.

Circular Permutations: The number of permutations of n distinct objects arranged in a circle is $(n-1)!$.

Permutations (all objects not distinct): The number of permutations of n objects of which n_1 are of one kind, n_2 are of a second kind, ... , n_k are of a kth kind, and $n_1 + n_2 + \cdots + n_k = n$, is

$$\frac{n!}{n_1! n_2! \cdots n_k!}$$

Combinations: The number of combinations of n distinct objects taken k at a time is

$$C_{n,k} = \binom{n}{k} = \frac{P_{n,k}}{k!} = \frac{n!}{k!(n-k)!}$$

1. For any positive integer n and $k = 0, 1, 2, \ldots, n$, $\dbinom{n}{k} = \dbinom{n}{n-k}$

2. For any positive integer n and $k = 1, 2, \ldots, n-1$, $\dbinom{n}{k} = \dbinom{n-1}{k} + \dbinom{n-1}{k-1}$

Partitions: The number of ways of partitioning a set of n distinct objects into k subsets with n_1 objects in the first subset, n_2 objects in the second subset, ... , and n_k objects in the kth subset, is

$$\binom{n}{n_1, n_2, \ldots, n_k} = \frac{n!}{n_1! n_2! \cdots n_k!}$$

Numerical Descriptive Statistics

The formulas in this section apply to a set of n observations $x_1, x_2, \ldots, x_n$.

Mean (Arithmetic Mean): $\bar{x} = \dfrac{1}{n} \sum\limits_{i=1}^{n} x_i = \dfrac{1}{n}(x_1 + x_2 + \cdots + x_n)$

Weighted Mean (Weighted Arithmetic Mean): Let $w_i > 0$ be the weight associated with x_i.

$$\bar{x} = \frac{\sum\limits_{i=1}^{n} w_i x_i}{\sum\limits_{i=1}^{n} w_i} = \frac{w_1 x_1 + w_2 x_2 + \cdots + w_n x_n}{w_1 + w_2 + \cdots + w_n}$$

Geometric Mean: $\text{GM} = \sqrt[n]{x_1 \cdot x_2 \cdots x_n}$, $x_i > 0$

Table 4. Probability and Statistics Formulas (Continued)

Harmonic Mean: $\text{HM} = \dfrac{n}{\sum\limits_{i=1}^{n} \frac{1}{x_i}} = \dfrac{n}{\frac{1}{x_1} + \frac{1}{x_2} + \cdots + \frac{1}{x_n}},$ $x_i > 0$

Relation Between Arithmetic, Geometric, and Harmonic Mean:
   $\text{HM} \leq \text{GM} \leq \bar{x},$ Equality holds if all the observations are equal.

$p\%$ Trimmed Mean: Eliminate the smallest $p\%$ and the largest $p\%$ of the sample. $\bar{x}_{\text{tr}(p)}$ is the arithmetic mean of the remaining data.

Mode: A mode of a set of n observations is a value which occurs most often, or with the greatest frequency. A mode may not exist and, if it exists, may not be unique.

Median: Rearrange the observations in increasing order,

$$\tilde{x} = \begin{cases} \text{the single middle value in the ordered list if } n \text{ is odd} \\ \text{the mean of the two middle values in the ordered list if } n \text{ is even} \end{cases}$$

Quartiles:
  1. $Q_2 = \tilde{x}$

  2. If n is even $\begin{cases} Q_1 \text{ is the median of the smallest } n/2 \text{ observations} \\ Q_3 \text{ is the median of the largest } n/2 \text{ observations} \end{cases}$

  3. If n is odd $\begin{cases} Q_1 \text{ is the median of the smallest } (n+1)/2 \text{ observations} \\ Q_3 \text{ is the median of the largest } (n+1)/2 \text{ observations} \end{cases}$

Mean Deviation: $\text{MD} = \dfrac{1}{n} \sum\limits_{i=1}^{n} |\, x_i - \bar{x}\,|$ or $\text{MD} = \dfrac{1}{n} \sum\limits_{i=1}^{n} |\, x_i - \tilde{x}\,|$

Variance: $s^2 = \dfrac{1}{n-1} \left[\sum\limits_{i=1}^{n} (x_i - \bar{x})^2 \right] = \dfrac{1}{n-1} \left[\sum\limits_{i=1}^{n} x_i^2 - \dfrac{1}{n} \left(\sum\limits_{i=1}^{n} x_i \right)^2 \right]$

Standard Deviation: $s = \sqrt{s^2}$

Standard Error of the Mean: $\text{SEM} = s/\sqrt{n}$

Root Mean Square: $\text{RMS} = \dfrac{1}{n} \sum\limits_{i=1}^{n} x_i^2$

Range: $R = \max\{x_1, x_2, \ldots, x_n\} - \min\{x_1, x_2, \ldots, x_n\} = x_{(n)} - x_{(1)}$

Lower Fourth: Q_1 *Upper Fourth:* Q_3 *Fourth Spread (Interquartile Range):* $f_s = IQR = Q_3 - Q_1$

Quartile Deviation (Semi-Interquartile Range): $(Q_3 - Q_1)/2$

Inner Fences: $Q_1 - 1.5f_s,\ Q_3 + 1.5f_s$ *Outer Fences:* $Q_1 - 3f_s,\ Q_3 + 3f_s$

Coefficient of Variation: $s/\bar{x}$

Coefficient of Quartile Variation: $(Q_3 - Q_1)/(Q_3 + Q_1)$

Moments:

   The rth moment about the origin: $m_r' = \dfrac{1}{n} \sum\limits_{i=1}^{n} x_i^r$

   The rth moment about the mean $\bar{x}$: $m_r = \dfrac{1}{n} \sum\limits_{i=1}^{n} (x_i - \bar{x})^r$

Table 4. Probability and Statistics Formulas (Continued)

Coefficient of Skewness: $g_1 = m_3/m_2^{3/2}$ *Coefficient of Kurtosis:* $g_2 = m_4/m_2^2$ *Coefficient of Excess:* $g_2 - 3$
where

$$m_2 = \frac{1}{n}\sum_{i=1}^{n}(x_i - \overline{x})^2 \qquad m_3 = \frac{1}{n}\sum_{i=1}^{n}(x_i - \overline{x})^3 \qquad m_4 = \frac{1}{n}\sum_{i=1}^{n}(x_i - \overline{x})^4$$

Data Transformations: Let $y_i = ax_i + b$, then $\overline{y} = a\overline{x} + b$ $s_y^2 = a^2 s_x^2$ $s_y = \mid a \mid s_x$

Probability

The sample space of an experiment, denoted S, is the set of all possible outcomes. Each element of a sample space is called an element of the sample space or a sample point. An event is any collection of outcomes contained in the sample space. A simple event consists of exactly one element and a compound event consists of more than one element.

Relative Frequency Concept of Probability: If an experiment is conducted n times in an identical and independent manner and $n(A)$ is the number of times the event A occurs, then $n(A)/n$ is the relative frequency of occurrence of the event A. As n increases, the relative frequency converges to a value called the limiting relative frequency of the event A. The probability of the event A occurring, $P(A)$, is this limiting relative frequency.

Axioms of Probability:
 1. For any event A, $P(A) \geq 0$.
 2. $P(S) = 1$.
 3. If $A_1, A_2, \ldots$, is a finite or infinite collection of pairwise mutually exclusive events of S, then

$$P(A_1 \cup A_2 \cup A_3 \cup \ldots) = P(A_1) + P(A_2) + P(A_3) + \ldots$$

The Probability Of An Event: The probability of an event A is the sum of $P(a_i)$ for all sample points a_i in the event A

$$P(A) = \sum_{a_i \in A} P(a_i)$$

Properties of Probability:
 1. If A and A' are complementary events, $P(A) = 1 - P(A')$.
 2. $P(\emptyset) = 0$ for any sample space S.
 3. For any events A and B, if $A \subset B$ then $P(A) \leq P(B)$.
 4. For any events A and B, $P(A \cup B) = P(A) + P(B) - P(A \cap B)$.
 5. If A and B are mutually exclusive events, then $P(A \cap B) = 0$.
 6. For any events A and B, $P(A) = P(A \cap B) + P(A \cap B')$
 7. For any events A, B, C,

$$P(A \cup B \cup C) = P(A) + P(B) + P(C) - P(A \cap B) - P(A \cap C) - P(B \cap C) + P(A \cap B \cap C)$$

 8. For any events $A_1, A_2, \ldots, A_n$,

$$P\left(\bigcup_{i=1}^{n} A_i\right) \leq \sum_{i=1}^{n} P(A_i) \qquad \text{Equality holds if the events are pairwise mutually exclusive.}$$

De Morgan's Laws: Let $A, A_1, A_2, \ldots, A_n$ and B be sets (events). Then
 1. $(A \cup B)' = A' \cap B'$

$$\left(\bigcup_{i=1}^{n} A_i\right)' = (A_1 \cup A_2 \cup \cdots \cup A_n)' = A_1' \cap A_2' \cap \cdots \cap A_n' = \bigcap_{i=1}^{n} A_i'$$

 2. $(A \cap B)' = A' \cup B'$

Table 4. Probability and Statistics Formulas (Continued)

$$\left(\bigcap_{i=1}^{n} A_i\right)' = (A_1 \cap A_2 \cap \cdots \cap A_n)' = A_1' \cup A_2' \cup \cdots \cup A_n' = \bigcup_{i=1}^{n} A_i'$$

Conditional Probability: The conditional probability of A given that B has occurred is

$$P(A \mid B) = \frac{P(A \cap B)}{P(B)}, \qquad P(B) > 0$$

1. If $P(A_1 \cap A_2 \cap \cdots \cap A_{n-1}) > 0$ then
   $P(A_1 \cap A_2 \cap \cdots \cap A_n) = P(A_1) \cdot P(A_2 \mid A_1) \cdot P(A_3 \mid A_1 \cap A_2) \cdots P(A_n \mid A_1 \cap A_2 \cap \cdots \cap A_{n-1})$
2. If $A \subset B$, then $P(A \mid B) = P(A)/P(B)$ and $P(B \mid A) = 1$
3. $P(A' \mid B) = 1 - P(A \mid B)$

The Multiplication Rule: $P(A \cap B) = P(A \mid B) \cdot P(B), \qquad P(B) \neq 0$

The Law of Total Probability: Let $A_1, A_2, \ldots, A_n$ be a collection of mutually exclusive, exhaustive events with $P(A_i) \neq 0$. Then for any event B,

$$P(B) = \sum_{i=1}^{n} P(B \mid A_i)P(A_i)$$

Bayes' Theorem: Let $A_1, A_2, \ldots, A_n$ be a collection of mutually exclusive exhaustive events, $P(A_i) \neq 0$. Then for any event B, $P(B) \neq 0$

$$P(A_k \mid B) = \frac{P(A_k \cap B)}{P(B)} = \frac{P(B \mid A_k)P(A_k)}{\sum_{i=1}^{n} P(B \mid A_i)P(A_i)}, \qquad k = 1, \ldots, n$$

Independence:
1. A and B are independent events if $P(A \mid B) = P(A)$, or equivalently if $P(B \mid A) = P(B)$.
2. A and B are independent events if and only if $P(A \cap B) = P(A) \cdot P(B)$.
3. $A_1, A_2, \ldots, A_n$ are pairwise independent events if $P(A_i \cap A_j) = P(A_i) \cdot P(A_j)$ for every pair i, j, $i \neq j$.
4. $A_1, A_2, \ldots, A_n$ are mutually independent events if for every k, $k = 2, 3, \ldots, n$, and every subset of indices $i_1, i_2, \ldots, i_k$, $P(A_{i_1} \cap A_{i_2} \cap \cdots \cap A_{i_k}) = P(A_{i_1}) \cdot P(A_{i_2}) \cdots P(A_{i_k})$.

Probability Distributions

Random Variable: Given a sample space S, a random variable is a function with domain S and range some subset of the real numbers. A random variable is discrete if it can assume only a finite or countably infinite number of values. A random variable is continuous if its set of possible values is an entire interval of numbers. Random variables will be denoted by upper-case letters, for example X.

Discrete Random Variables

Probability Mass Function: The probability distribution or probability mass function (pmf) of a discrete random variable is defined for every number x by $p(x) = P(X = x)$.

1. $p(x) \geq 0$
2. $\sum_x p(x) = 1$

Cumulative Distribution Function: The cumulative distribution function (cdf) $F(x)$ of a discrete random variable X with pmf $p(x)$ is defined for every number x by

$$F(x) = P(X \leq x) = \sum_{y:y \leq x} p(y)$$

Table 4. Probability and Statistics Formulas (Continued)

1. $\lim_{x \to -\infty} F(x) = 0$

2. $\lim_{x \to \infty} F(x) = 1$

3. For any real numbers a and b, if $a < b$, then $F(a) \leq F(b)$.

Continuous Random Variables

Probability Density Function: The probability distribution or probability density function (pdf) of a continuous random variable X is a function $f(x)$ such that

$$P(a \leq X \leq b) = \int_a^b f(x)\,dx, \quad a, b \in \Re \text{ the set of reals}, \quad a \leq b$$

1. $f(x) \geq 0 \quad$ for $\quad -\infty < x < \infty$

2. $\int_{-\infty}^{\infty} f(x)\,dx = 1$

3. $P(X = c) = 0 \quad$ for $\quad c \in \Re$.

4. $P(a \leq X \leq b) = P(a < X \leq b) = P(a \leq X < b) = P(a < X < b)$ for $a, b \in \Re$ and $a < b$.

Cumulative Distribution Function: The cumulative distribution function (cdf) $F(x)$ for a continuous random variable X is defined by

$$F(x) = P(X \leq x) = \int_{-\infty}^{x} f(y)\,dy, \quad -\infty < x < \infty$$

1. $P(a \leq X \leq b) = P(X \leq b) - P(X \leq a) = F(b) - F(a), \quad a, b \in \Re$ and $a < b$.

2. The pdf $f(x)$ can be found from the cdf:

$$f(x) = \frac{dF(x)}{dx} \quad \text{whenever the derivative exists}$$

Mathematical Expectation

Expected Value:
1. If X is a discrete random variable with pmf $p(x)$,

    a. the expected value of X is $E(X) = \mu = \sum_x x p(x)$.

    b. the expected value of a function $g(X)$ is $E[g(X)] = \mu_{g(X)} = \sum_x g(x) p(x)$.

2. If X is a continuous random variable with pdf $f(x)$,

    a. the expected value of X is $E(X) = \mu = \int_{-\infty}^{\infty} x f(x)\,dx$.

    b. the expected value of a function $g(X)$ is $E[g(X)] = \mu_{g(X)} = \int_{-\infty}^{\infty} g(x) f(x)\,dx$.

Theorems:
1. $E(aX + bY) = aE(X) + bE(Y)$
2. $E(X \cdot Y) = E(X) \cdot E(Y)$ if X and Y are independent.

Variance: The variance of a random variable X is

$$\sigma^2 = \mathrm{Var}(X) = E[(X - \mu)^2] = \begin{cases} \sum_x (x - \mu)^2 p(x) & \text{if } X \text{ is discrete} \\ \int_{-\infty}^{\infty} (x - \mu)^2 f(x)\,dx & \text{if } X \text{ is continuous} \end{cases}$$

Table 4. Probability and Statistics Formulas (Continued)

The *standard deviation* of X is $\sigma = \sqrt{\sigma^2}$

Theorems:

1. $\sigma^2 = E(X^2) - [E(X)]^2$

2. $\sigma_{aX}^2 = a^2 \cdot \sigma_X^2 \qquad \sigma_{aX} = |a| \cdot \sigma_X$

3. $\sigma_{X+b}^2 = \sigma_X^2$

4. $\sigma_{aX+b}^2 = a^2 \cdot \sigma_X^2 \qquad \sigma_{aX+b} = |a| \cdot \sigma_X$

Chebyshev's Theorem: Let X be a random variable with mean μ and variance σ^2. For any constant $k > 0$

$$P(|X - \mu| < k\sigma) = P(\mu - k\sigma < X < \mu + k\sigma) \geq 1 - \frac{1}{k^2}$$

Jensen's Inequality: Let $h(x)$ be a function such that $\frac{d^2}{dx^2} h(x) \geq 0$ then $E[h(X)] \geq h(E[X])$.

Moments About The Origin: The rth moment about the origin, $r = 0, 1, 2, \ldots$, of a random variable X is

$$\mu_r' = E(X^r) = \begin{cases} \sum_x x^r p(x) & \text{if } X \text{ is discrete} \\ \int_{-\infty}^{\infty} x^r f(x)\, dx & \text{if } X \text{ is continuous} \end{cases}$$

In particular $\mu_1' = E(X) = \mu$

Moments About The Mean: The rth moment about the mean, $r = 0, 1, 2, \ldots$, of a random variable X is

$$\mu_r = E[(X - \mu)^r] = \begin{cases} \sum_x (x - \mu)^r p(x) & \text{if } X \text{ is discrete} \\ \int_{-\infty}^{\infty} (x - \mu)^r f(x)\, dx & \text{if } X \text{ is continuous} \end{cases}$$

In particular $\mu_2 = E[(X - \mu)^2] = \sigma^2$

Factorial Moments: The rth factorial moment, $r = 1, 2, 3, \ldots$, of a random variable X is

$$\mu_{[r]} = E[X(X-1)(X-2) \cdots (X-r+1)]$$

$$= \begin{cases} \sum_x x(x-1)(x-2) \cdots (x-r+1)p(x) & \text{if } X \text{ is discrete} \\ \int_{-\infty}^{\infty} x(x-1)(x-2) \cdots (x-r+1)f(x)\, dx & \text{if } X \text{ is continuous} \end{cases}$$

Moment-generating Functions: The moment-generating function (mgf) of a random variable X, where it exists, is

$$M(t) = E\left(e^{tX}\right) = \begin{cases} \sum_x e^{tx} p(x) & \text{if } X \text{ is discrete} \\ \int_{-\infty}^{\infty} e^{tx} f(x)\, dx & \text{if } X \text{ is continuous} \end{cases}$$

Theorems:

1. If $M(t)$ exists, then for $r = 1, 2, \ldots$ $\qquad \mu_r' = M^{(r)}(0) = \left. \dfrac{d^r M(t)}{dt^r} \right|_{t=0}$

2. $M_{aX}(t) = M_X(at)$

3. $M_{X+b}(t) = e^{bt} \cdot M_X(t)$

4. $M_{(X+b)/a}(t) = e^{(b/a)t} \cdot M_X(t/a)$

Probability-generating Function: The probability-generating function for a discrete random variable X is

Table 4. Probability and Statistics Formulas (Continued)

$$P(t) = E\left(t^X\right) = \sum_x t^x p(x)$$

Theorem: $\mu_{[r]} = P^{(r)}(1) = \dfrac{d^r P(t)}{dt^r}\bigg|_{t=1}$

Characteristic Function: The characteristic function of a random variable X is

$$\phi(t) = E\left(e^{itX}\right) = \begin{cases} \sum_x e^{itx} p(x) & \text{if } X \text{ is discrete} \\ \int_{-\infty}^{\infty} e^{itx} f(x)\, dx & \text{if } X \text{ is continuous} \end{cases}$$

where t is a real number and $i^2 = -1$.

Theorem: $i^r \mu_r' = \phi^{(r)}(0) = \dfrac{d^r \phi(t)}{dt^r}\bigg|_{t=0}$

Multivariate Distributions

Discrete Case: Let X and Y be discrete random variables. The joint (bivariate) probability distribution or joint probability mass function for X and Y is

$$p(x, y) = P(X = x, Y = y) \qquad \forall\, (x, y)$$

1. For any set A consisting of pairs (x, y), $P[(X, Y) \in A] = \sum\sum_{(x,y) \in A} p(x, y)$

2. $p(x, y) \geq 0 \quad \forall\, (x, y)$

3. $\sum_x \sum_y p(x, y) = 1$

Continuous Case: Let X and Y be continuous random variables. Then $f(x, y)$ is the joint probability density function for X and Y if for any two-dimensional set A

$$P[(X, Y) \in A] = \int\int_A f(x, y)\, dx\, dy$$

1. If A is a rectangle $\{(x, y) \mid a \leq x \leq b, c \leq y \leq d\}$, then

$$P[(X, Y) \in A] = P(a \leq X \leq b, c \leq Y \leq d) = \int_a^b \int_c^d f(x, y)\, dy\, dx$$

2. $f(x, y) \geq 0 \quad \forall\, (x, y)$

3. $\int_{-\infty}^{\infty} \int_{-\infty}^{\infty} f(x, y)\, dx\, dy = 1$

Joint Distribution Function: For any two random variables X and Y the joint distribution function is $F(x, y) = P(X \leq x, Y \leq y)$.

$$F(a, b) = \begin{cases} \sum_{x=-\infty}^{a} \sum_{y=-\infty}^{b} p(x, y) & \text{if } X \text{ and } Y \text{ are discrete} \\ \int_{-\infty}^{a} \int_{-\infty}^{b} f(x, y)\, dy\, dx & \text{if } X \text{ and } Y \text{ are continuous} \end{cases}$$

Properties:

1. $\lim\limits_{(x,y) \to (-\infty, -\infty)} F(x, y) = \lim\limits_{x \to -\infty} F(x, y) = \lim\limits_{y \to -\infty} F(x, y) = 0$

2. $\lim\limits_{(x,y) \to (\infty, \infty)} F(x, y) = 1$

3. If $a \leq b$ and $c \leq d$, then

Table 4. Probability and Statistics Formulas (Continued)

$$P(a < X \le b, c < Y \le d) = F(b, d) - F(b, c) - F(a, d) + F(a, c) \ge 0$$

4. The joint probability density function can be found from the joint distribution function:

$$f(x, y) = \frac{\partial^2}{\partial x \partial y} F(x, y) \quad \text{whenever the partials exist.}$$

n Random Variables

Discrete Case: Let $X_1, X_2, \ldots, X_n$ be discrete random variables. The joint distribution for $X_1, X_2, \ldots, X_n$ is

$$p(x_1, x_2, \ldots, x_n) = P(X_1 = x_1, X_2 = x_2, \ldots, X_n = x_n) \quad \forall \ (x_1, x_2, \ldots, x_n)$$

For any set A consisting of n-tuples $(x_1, x_2, \ldots, x_n)$

$$P[(X_1, X_2, \ldots, X_n) \in A] = \sum \sum \cdots \sum_{(x_1, x_2, \ldots, x_n) \in A} p(x_1, x_2, \ldots, x_n)$$

Continuous Case: Let $X_1, X_2, \ldots, X_n$ be continuous random variables. Then $f(x_1, x_2, \ldots, x_n)$ is the joint probability density function for $X_1, X_2, \ldots, X_n$ if for any set A

$$P[(X_1, X_2, \ldots, X_n) \in A] = \int \int \cdots \int_A f(x_1, x_2, \ldots, x_n) \, dx_1 \, dx_2 \cdots dx_n$$

Joint Distribution Function: The joint distribution function for the n random variables $X_1, X_2, \ldots, X_n$ is

$$F(x_1, x_2, \ldots, x_n) = P(X_1 \le x_1, X_2 \le x_2, \ldots, X_n \le x_n)$$

$$F(y_1, y_2, \ldots, y_n) = \begin{cases} \sum_{x_1 = -\infty}^{y_1} \sum_{x_2 = -\infty}^{y_2} \cdots \sum_{x_n = -\infty}^{y_n} p(x_1, x_2, \ldots, x_n) & \text{if } X_1, X_2, \ldots, X_n \text{ are discrete} \\ \int_{-\infty}^{y_1} \int_{-\infty}^{y_2} \cdots \int_{-\infty}^{y_n} f(x_1, x_2, \ldots, x_n) \, dx_n \cdots dx_1 & \text{if } X_1, X_2, \ldots, X_n \text{ are continuous} \end{cases}$$

$$f(x_1, x_2, \ldots, x_n) = \frac{\partial^n}{\partial x_1 \partial x_2 \cdots \partial x_n} F(x_1, x_2, \ldots, x_n) \quad \text{whenever the partials exist.}$$

Marginal Distributions

1. Let X and Y be discrete random variables. The marginal probability mass functions for X and Y are

$$p_X(x) = \sum_y p(x, y) \qquad p_Y(y) = \sum_x p(x, y)$$

2. Let X and Y be continuous random variables. The marginal probability density functions for X and Y are

$$f_X(x) = \int_{-\infty}^{\infty} f(x, y) \, dy \quad \text{for} \quad -\infty < x < \infty \qquad f_Y(y) = \int_{-\infty}^{\infty} f(x, y) \, dx \quad \text{for} \quad -\infty < y < \infty$$

3. Let $X_1, X_2, \ldots, X_n$ be a collection of random variables. The marginal distribution of a subset of the random variables, $X_1, X_2, \ldots, X_r$ $(r < n)$ is

$$g(x_1, x_2, \ldots x_r) = \begin{cases} \sum_{x_{r+1}} \cdots \sum_{x_n} p(x_1, x_2, \ldots, x_n) & \text{if } X_1, X_2, \ldots, X_n \text{ are discrete} \\ \int_{-\infty}^{\infty} \cdots \int_{-\infty}^{\infty} f(x_1, x_2, \ldots, x_n) \, dx_{r+1} \, dx_{r+2} \cdots dx_n & \text{if } X_1, X_2, \ldots, X_n \text{ are continuous} \end{cases}$$

Conditional Distributions

1. Let X and Y be discrete random variables with joint probability mass function $p(x, y)$ and let $p_Y(y)$ be the marginal probability mass function for Y. The conditional probability mass function for X given $Y = y$ is

$$p(x \mid y) = \frac{p(x, y)}{p_Y(y)}, \quad p_Y(y) \ne 0$$

Table 4. Probability and Statistics Formulas (Continued)

Let $p_X(x)$ be the marginal probability mass function for X. The conditional probability mass function for Y given $X = x$ is

$$p(y \mid x) = \frac{p(x, y)}{p_X(x)}, \quad p_X(x) \neq 0$$

2. Let X and Y be continuous random variables with joint probability density function $f(x, y)$ and let $f_Y(y)$ be the marginal probability density function for Y. The conditional probability density function for X given $Y = y$ is

$$f(x \mid y) = \frac{f(x, y)}{f_Y(y)}, \quad f_Y(y) \neq 0$$

Let $f_X(x)$ be the marginal probability density function for X. The conditional probability density function for Y given $X = x$ is

$$f(y \mid x) = \frac{f(x, y)}{f_X(x)}, \quad f_X(x) \neq 0$$

3. Let $X_1, X_2, \ldots, X_n$ be a collection of random variables. The conditional distribution of any subset $X_1, X_2, \ldots, X_k$ given $X_{k+1} = x_{k+1}, X_{k+2} = x_{k+2}, \ldots, X_n = x_n$ is

$$p(x_1, x_2, \ldots, x_k \mid x_{k+1}, x_{k+2}, \ldots, x_k) = \frac{p(x_1, x_2, \ldots, x_n)}{g(x_{k+1}, x_{k+2}, \ldots, x_n)}, \quad g(x_{k+1}, x_{k+2}, \ldots, x_n) \neq 0$$

if $X_1, X_2, \ldots, X_n$ are discrete with joint probability mass function $p(x_1, x_2, \ldots, x_n)$ and the random variables $X_{k+1}, X_{k+2}, \ldots, X_n$ have marginal probability mass function $g(x_{k+1}, x_{k+2}, \ldots, x_n)$,

$$f(x_1, x_2, \ldots, x_k \mid x_{k+1}, x_{k+2}, \ldots, x_k) = \frac{f(x_1, x_2, \ldots, x_n)}{g(x_{k+1}, x_{k+2}, \ldots, x_n)}, \quad g(x_{k+1}, x_{k+2}, \ldots, x_n) \neq 0$$

if $X_1, X_2, \ldots, X_n$ are continuous with joint probability density function $f(x_1, x_2, \ldots, x_n)$ and the random variables $X_{k+1}, X_{k+2}, \ldots, X_n$ have marginal probability density function $g(x_{k+1}, x_{k+2}, \ldots, x_n)$.

Independent Random Variables: Let $X_1, X_2, \ldots, X_n$ be a collection of discrete (continuous) random variables with joint probability mass (density) function $p(x_1, x_2, \ldots, x_n)$ ($f(x_1, x_2, \ldots, x_n)$). Let $p_{X_i}(x_i)$ ($f_{X_i}(x_i)$) be the marginal probability mass (density) function for X_i for $i = 1, 2, \ldots, n$. The random variables are independent if and only if

$$p(x_1, x_2, \ldots, x_n) = p_{X_1}(x_1) \cdot p_{X_2}(x_2) \cdots p_{X_n}(x_n)$$
$$(f(x_1, x_2, \ldots, x_n) = f_{X_1}(x_1) \cdot f_{X_2}(x_2) \cdots f_{X_n}(x_n))$$

for all $x_1, x_2, \ldots, x_n$.

The Expected Value of a Function of Random Variables: Let $g(X_1, X_2, \ldots, X_n)$ be a function of the random variables $X_1, X_2, \ldots, X_n$. If $X_1, X_2, \ldots, X_n$ are discrete random variables with joint probability mass function $p(x_1, x_2, \ldots, x_n)$ then the expected value of $g(X_1, X_2, \ldots, X_n)$ is

$$E[g(X_1, X_2, \ldots, X_n)] = \sum_{x_1} \sum_{x_2} \cdots \sum_{x_n} g(x_1, x_2, \ldots, x_n) p(x_1, x_2, \ldots, x_n)$$

If $X_1, X_2, \ldots, X_n$ are continuous random variables with joint density function $f(x_1, x_2, \ldots, x_n)$ then

$$E[g(X_1, X_2, \ldots, X_n)] = \int_{-\infty}^{\infty} \int_{-\infty}^{\infty} \cdots \int_{-\infty}^{\infty} g(x_1, x_2, \ldots, x_n) f(x_1, x_2, \ldots, x_n) \, dx_1 \, dx_2 \cdots dx_n$$

Theorem: Let $c_1, c_2, \ldots, c_n$ be constants, then

$$E\left[\sum_{i=1}^{n} c_i g_i(X_1, X_2, \ldots, X_n) \right] = \sum_{i=1}^{n} c_i E[g_i(X_1, X_2, \ldots, X_n)]$$

The Product Moment About The Origin: The rth and sth product moment about the origin of the random

Table 4. Probability and Statistics Formulas (Continued)

variables X and Y is defined for $r = 0, 1, 2, \ldots$, and $s = 0, 1, 2, \ldots$, by

$$\mu'_{r,s} = E(X^r Y^s) = \begin{cases} \sum_x \sum_y x^r y^s p(x, y) & \text{if } X \text{ and } Y \text{ are discrete} \\ \int_{-\infty}^{\infty} \int_{-\infty}^{\infty} x^r y^s f(x, y)\, dx\, dy & \text{if } X \text{ and } Y \text{ are continuous} \end{cases}$$

The Product Moment About The Means: The rth and sth product moment about the respective means of the random variables X and Y is defined for $r = 0, 1, 2, \ldots$, and $s = 0, 1, 2, \ldots$, by

$$\mu_{r,s} = E[(X - \mu_X)^r (Y - \mu_Y)^s] = \begin{cases} \sum_x \sum_y (x - \mu_X)^r (y - \mu_Y)^s p(x, y) & \text{if } X \text{ and } Y \text{ are discrete} \\ \int_{-\infty}^{\infty} \int_{-\infty}^{\infty} (x - \mu_X)^r (y - \mu_Y)^s f(x, y)\, dx\, dy & \text{if } X \text{ and } Y \text{ are continuous} \end{cases}$$

Covariance: The covariance of the random variables X and Y is defined to be

$$\sigma_{XY} = \text{Cov}(X, Y) = \mu_{1,1} = E[(X - \mu_x)(Y - \mu_Y)]$$

Theorems:

1. If $X_1, X_2, \ldots, X_n$ are independent, then

$$E(X_1 X_2 \cdots X_n) = E(X_1) E(X_2) \cdots E(X_n)$$

2. $\text{Cov}(X, Y) = \mu'_{1,1} - \mu_x \mu_y = E(XY) - E(X)E(Y)$

3. If X and Y are independent random variables, then $\text{Cov}(X, Y) = 0$.

Linear Combinations Of Random Variables

Let $X_1, X_2, \ldots, X_m$ and $Y_1, Y_2, \ldots, Y_n$ be random variables and $a_1, a_2, \ldots, a_m$ and $b_1, b_2, \ldots, b_n$ be constants. Define

$$U = \sum_{i=1}^{m} a_i X_i \qquad V = \sum_{j=1}^{n} b_j Y_j$$

Theorems:

1. $E(U) = \sum_{i=1}^{m} a_i E(X_i)$

2. $\text{Var}(U) = \sum_{i=1}^{m} a_i^2 \text{Var}(X_i) + 2 \sum \sum_{i<j} a_i a_j \text{Cov}(X_i, X_j),$

    where the double sum extends over all pairs (i, j) with $i < j$.

3. If the random variables $X_1, X_2, \ldots, X_m$ are independent, $\text{Var}(U) = \sum_{i=1}^{m} a_i^2 \text{Var}(X_i)$.

4. $\text{Cov}(U, V) = \sum_{i=1}^{m} \sum_{j=1}^{n} a_i b_j \text{Cov}(X_i, Y_j)$

Conditional Expectation: Let X and Y be random variables and let $g(X)$ be a function of X. The conditional expectation of $g(X)$ given $Y = y$ is defined by

$$E[g(X) \mid y] = \begin{cases} \sum_x g(x) p(x \mid y) & \text{if } X \text{ and } Y \text{ are discrete} \\ \int_{-\infty}^{\infty} g(x) f(x \mid y)\, dx & \text{if } X \text{ and } Y \text{ are continuous} \end{cases}$$

1. The conditional mean, or conditional expectation, of X given $Y = y$ is

$$\mu_{X|y} = E(X \mid y) = \begin{cases} \sum_x x p(x \mid y) & \text{if } X \text{ and } Y \text{ are discrete} \\ \int_{-\infty}^{\infty} x f(x \mid y)\, dx & \text{if } X \text{ and } Y \text{ are continuous} \end{cases}$$

2. The conditional variance of X given $Y = y$ is

Table 4. Probability and Statistics Formulas (Continued)

$$\sigma^2_{X|y} = E[(X - \mu_{X|y})^2 \mid y] = E(X^2 \mid y) - \mu^2_{X|y}$$

3. $E(X) = E[E(X \mid Y)]$

Special Distributions

The Multinomial Distribution: The random variables $X_1, X_2, \ldots, X_n$ have a multinomial distribution if their joint probability distribution is given by

$$p(x_1, x_2, \ldots, x_n) = \binom{n}{x_1, x_2, \ldots, x_n} p_1^{x_1} p_2^{x_2} \cdots p_n^{x_n}$$

for $x_i = 0, 1, \ldots, n$ for each i and $\sum_{i=1}^{n} x_i = n$, $\sum_{i=1}^{n} p_i = 1$.

1. $E(X_i) = np_i$

2. $\text{Var}(X_i) = np_i(1 - p_i)$

3. $\text{Cov}(X_i, X_j) = -np_i p_j$, $\quad i \neq j$

The Bivariate Normal Distribution: The random variables X and Y have a bivariate normal distribution if their joint probability density function is given by

$$f(x, y) = \frac{e^{-\frac{1}{2(1-\rho^2)}\left[\left(\frac{x-\mu_x}{\sigma_X}\right)^2 - 2\rho\left(\frac{x-\mu_X}{\sigma_X}\right)\left(\frac{y-\mu_Y}{\sigma_Y}\right) + \left(\frac{y-\mu_Y}{\sigma_Y}\right)^2\right]}}{2\pi\sigma_X\sigma_Y\sqrt{1-\rho^2}}, \quad -\infty < x < \infty, \quad -\infty < y < \infty$$

where $\sigma_X > 0$, $\sigma_Y > 0$, and $-1 < \rho < 1$.

Theorems:

1. $E(X) = \mu_X$, $E(Y) = \mu_Y$, $\text{Var}(X) = \sigma_X^2$, $\text{Var}(Y) = \sigma_Y^2$, $\text{Cov}(X, Y) = \rho\sigma_X\sigma_Y$

2. The conditional density of X given $Y = y$ is a normal distribution with

$$\mu_{X|y} = \mu_X + \rho\frac{\sigma_X}{\sigma_Y}(y - \mu_Y) \quad \text{and} \quad \sigma^2_{X|y} = \sigma_X^2(1 - \rho^2)$$

The conditional density of Y given $X = x$ is a normal distribution with

$$\mu_{Y|x} = \mu_Y + \rho\frac{\sigma_Y}{\sigma_X}(x - \mu_X) \quad \text{and} \quad \sigma^2_{Y|x} = \sigma_Y^2(1 - \rho^2)$$

3. X and Y are independent if and only if $\rho = 0$.

Functions of Random Variables

Given a collection of random variables $X_1, X_2, \ldots, X_n$ and their joint probability mass function or joint probability density function, let the random variable $Y = Y(X_1, X_2, \ldots, X_n)$ be a function of $X_1, X_2, \ldots, X_n$. The following are techniques for determining the probability distribution of Y.

Method of Distribution Functions:
1. Determine the region $Y = y$ in the $(x_1, x_2, \ldots, x_n)$ space.
2. Determine the region $Y \leq y$.
3. Compute $F(y) = P(Y \leq y)$ by integrating the joint probability density function $f(x_1, x_2, \ldots, x_n)$ over the region $Y \leq y$.
4. Compute the probability density function for Y, $f(y)$, by differentiating $F(y)$, that is

$$f(y) = \frac{dF(y)}{dy}$$

Method of Transformations (One Variable): Let X be a random variable with probability density function $f_X(x)$. If $u(x)$ is differentiable and either increasing or decreasing, then $Y = u(X)$ has probability

19

Table 4. Probability and Statistics Formulas (Continued)

density function

$$f_Y(y) = f_X(w(y)) \cdot \mid w'(y) \mid, \quad u'(x) \neq 0$$

where $x = w(y) = u^{-1}(y)$

Method of Transformations (Two Variables): Let X_1 and X_2 be random variables with joint probability density function $f(x_1, x_2)$. Let the functions $y_1 = u_1(x_1, x_2)$ and $y_2 = u_2(x_1, x_2)$ represent a one-to-one transformation from the x's to the y's and let the partial derivatives with respect to both x_1 and x_2 exist. Then the joint probability density function for $Y_1 = u_1(X_1, X_2)$ and $Y_2 = u_2(X_1, X_2)$ is

$$g(y_1, y_2) = f(w_1(y_1, y_2), w_2(y_1, y_2)) \cdot \mid J \mid$$

where $y_1 = u_1(x_1, x_2)$ and $y_2 = u_2(x_1, x_2)$ are uniquely solved for $x_1 = w_1(y_1, y_2)$ and $x_2 = w_2(y_1, y_2)$, and J is the determinant of the Jacobian

$$J = \begin{vmatrix} \frac{\partial x_1}{\partial y_1} & \frac{\partial x_1}{\partial y_2} \\ \frac{\partial x_2}{\partial y_1} & \frac{\partial x_2}{\partial y_2} \end{vmatrix}$$

Method of Moment-generating Functions: Let Y be a function of the random variables $X_1, X_2, \ldots, X_n$.
  1. Determine the moment-generating function for Y, $M_Y(t)$.
  2. If $M_Y(t) = M_U(t)$ for all t, then Y and U have identical distributions.

Theorems:

  1. Let X be a random variable with moment-generating function $M_X(t)$ and let Y be a random variable with moment-generating function $M_Y(t)$. If $M_X(t) = M_Y(t)$ for all t, then X and Y have the same probability distribution.

  2. Let $X_1, X_2, \ldots, X_n$ be independent random variables and let $Y = X_1 + X_2 + \cdots + X_n$, then

$$M_Y(t) = \prod_{i=1}^{n} M_{X_i}(t)$$

Sampling Distributions

Definitions:
  1. The random variables $X_1, X_2, \ldots, X_n$ are said to be a random sample of size n from an infinite population if $X_1, X_2, \ldots, X_n$ are independent and identically distributed (iid).
  2. Let $X_1, X_2, \ldots, X_n$ be a random sample, the sample mean is

$$\overline{X} = \frac{1}{n} \sum_{i=1}^{n} X_i$$

The sample variance is

$$S^2 = \frac{1}{n-1} \sum_{i=1}^{n} (X_i - \overline{X})^2$$

Theorem: Let $X_1, X_2, \ldots, X_n$ be a random sample from an infinite population with mean μ and variance σ^2, then

$$E(\overline{X}) = \mu \quad \text{and} \quad \text{Var}(\overline{X}) = \frac{\sigma^2}{n}$$

The Standard Error of the Mean: $\sigma_{\overline{X}} = \dfrac{\sigma}{\sqrt{n}}$

The Law of Large Numbers: For any positive constant c, $P(\mu - c < \overline{X} < \mu + c) \geq 1 - \frac{\sigma^2}{nc^2}$.

The Central Limit Theorem: Let $X_1, X_2, \ldots, X_n$ be a random sample from an infinite population with mean

20

Table 4. Probability and Statistics Formulas (Continued)

μ and variance σ^2. The limiting distribution of

$$Z = \frac{\overline{X} - \mu}{\sigma/\sqrt{n}}$$

as $n \to \infty$ is the standard normal distribution.

Theorem: Let $X_1, X_2, \ldots, X_n$ be a random sample from a normal population with mean μ and variance σ^2. Then $\overline{X}$ is normally distributed with mean μ and variance σ^2/n.

The Distribution of The Mean: Finite Populations

Let $\{c_1, c_2, \ldots, c_N\}$ be a collection of numbers representing a finite population of size N and assume the sampling from this population is done without replacement. Let the random variable X_i be the ith observation from the population. Then $X_1, X_2, \ldots, X_n$ is a random sample from a finite population if the joint probability mass function of $X_1, X_2, \ldots, X_n$ is

$$p(x_1, x_2, \ldots, x_n) = \frac{1}{N(N-1)\cdots(N-n+1)}$$

1. The marginal distribution of the random variable X_i, $i = 1, 2, \ldots, n$, is

$$p_{X_i}(x_i) = \frac{1}{N} \quad \text{for} \quad x_i = c_1, c_2, \ldots, c_n$$

2. The mean and variance of the finite population $c_1, c_2, \ldots, c_n$ are

$$\mu = \sum_{i=1}^{N} c_i \frac{1}{N} \quad \text{and} \quad \sigma^2 = \sum_{i=1}^{N} (c_i - \mu)^2 \frac{1}{N}$$

3. The joint marginal probability mass function of any two of the random variables $X_1, X_2, \ldots, X_n$ is

$$p(x_i, x_j) = \frac{1}{N(N-1)}$$

4. The covariance between any two of the random variables $X_1, X_2, \ldots, X_n$ is

$$\text{Cov}(X_i, X_j) = -\frac{\sigma^2}{N-1}$$

5. Let $\overline{X}$ be the sample mean of the random sample of size n. Then

$$E(\overline{X}) = \mu \quad \text{and} \quad \text{Var}(\overline{X}) = \frac{\sigma^2}{n} \cdot \frac{N-n}{N-1}$$

The quantity $(N-n)/(N-1)$ is the finite population correction factor.

The Chi-Square Distribution, Theorems:

1. Let Z be a standard normal random variable, then Z^2 has a chi-square distribution with 1 degree of freedom.

2. Let $Z_1, Z_2, \ldots, Z_n$ be independent standard normal random variables, then

$$Y = \sum_{i=1}^{n} Z_i^2$$

has a chi-square distribution with n degrees of freedom.

3. Let $X_1, X_2, \ldots, X_n$ be independent random variables such that X_i has a chi-square distribution with ν_i degrees of freedom. Then

$$Y = \sum_{i=1}^{n} X_i$$

Table 4. Probability and Statistics Formulas (Continued)

has a chi-square distribution with $\nu = \nu_1 + \nu_2 + \cdots + \nu_n$ degrees of freedom.

4. Let U have a chi-square distribution with ν_1 degrees of freedom, U and V be independent, and $U + V$ have a chi-square distribution with $\nu > \nu_1$ degrees of freedom. Then V has a chi-square distribution with $\nu - \nu_1$ degrees of freedom.

5. Let $X_1, X_2, \ldots, X_n$ be a random sample from a normal population with mean μ and variance σ^2. Then

   a. $\overline{X}$ and S^2 are independent, and

   b. the random variable $\frac{(n-1)S^2}{\sigma^2}$ has a chi-square distribution with $n - 1$ degrees of freedom.

The t Distribution, Theorems:

1. Let Z have a standard normal distribution, X have a chi-square distribution with ν degrees of freedom, and X and Z be independent. Then

$$T = \frac{Z}{\sqrt{X/\nu}}$$

has a t distribution with ν degrees of freedom.

2. Let $X_1, X_2, \ldots, X_n$ be a random sample from a normal population with mean μ and variance σ^2. Then

$$T = \frac{\overline{X} - \mu}{S/\sqrt{n}}$$

has a t distribution with $n - 1$ degrees of freedom.

The F Distribution, Theorems:

1. Let U have a chi-square distribution with ν_1 degrees of freedom, V have a chi-square distribution with ν_2 degrees of freedom, and U and V be independent. Then

$$F = \frac{U/\nu_1}{V/\nu_2}$$

has and F distribution with ν_1 and ν_2 degrees of freedom.

2. Let $X_1, X_2, \ldots, X_m$ and $Y_1, Y_2, \ldots, Y_n$ be random samples from normal populations with variances σ_X^2 and σ_Y^2, respectively. Then

$$F = \frac{S_X^2/\sigma_X^2}{S_Y^2/\sigma_Y^2}$$

has a F distribution with $m - 1$ and $n - 1$ degrees of freedom.

3. Let F_{α,ν_1,ν_2} be critical value for the F distribution defined by $P(F \geq F_{\alpha,\nu_1,\nu_2}) = \alpha$. Then $F_{1-\alpha,\nu_1,\nu_2} = 1/F_{\alpha,\nu_2,\nu_1}$

Order Statistics

Definition: Let $X_1, X_2, \ldots, X_n$ be independent continuous random variables with probability density function $f(x)$ and cumulative distribution function $F(x)$. The order statistic, $X_{(i)}$, $i = 1, 2, \ldots, n$, is a random variable defined to be the ith largest of the set $\{X_1, X_2, \ldots, X_n\}$. Thus

$$X_{(1)} \leq X_{(2)} \leq \cdots \leq X_{(n)}$$

and in particular

$$X_{(1)} = \min\{X_1, X_2, \ldots, X_n\} \quad \text{and} \quad X_{(n)} = \max\{X_1, X_2, \ldots, X_n\}$$

The joint density of $X_1, X_2, \ldots, X_n$ is

$$g(x_1, x_2, \ldots, x_n) = n! f(x_1) \cdot f(x_2) \cdots f(x_n)$$

Table 4. Probability and Statistics Formulas (Continued)

The First Order Statistic: The probability density function, $g_1(x)$, and the cumulative distribution function, $G_1(x)$, for $X_{(1)}$ are

$$g_1(x) = n[1 - F(x)]^{n-1} f(x) \qquad G_1(x) = 1 - [1 - F(x)]^n$$

The nth Order Statistic: The probability density function, $g_n(x)$, and the cumulative distribution function, $G_n(x)$, for $X_{(n)}$ are

$$g_n(x) = n[F(x)]^{n-1} f(x) \qquad G_n(x) = [F(x)]^n$$

The ith Order Statistic: The probability density function, $g_i(x)$, for the ith order statistic is

$$g_i(x) = \frac{n!}{(i-1)!(n-i)!} [F(x)]^{i-1} f(x) [1 - F(x)]^{n-i}$$

Estimation

Let $\hat{\theta}$ be a point estimator of the parameter θ.

Unbiased Estimator: $\hat{\theta}$ is an unbiased estimator of θ if $E(\hat{\theta}) = \theta$.

Bias: The bias of $\hat{\theta}$ is $B(\hat{\theta}) = E(\hat{\theta}) - \theta$.

Mean Square Error: The mean square error of $\hat{\theta}$ is $MSE(\hat{\theta}) = E[(\hat{\theta} - \theta)^2] = \text{Var}(\hat{\theta}) + B(\hat{\theta})^2$.

Error of Estimation: The error of estimation is $\epsilon = |\hat{\theta} - \theta|$.

Cramér-Rao Inequality: Let $X_1, X_2, \ldots, X_n$ be a random sample from a population with probability density function $f(x)$. Let $\hat{\theta}$ be an unbiased estimator of θ. Under very general conditions it can be shown that

$$\text{Var}(\hat{\theta}) \geq \frac{1}{n \cdot E\left[\left(\frac{\partial \ln f(X)}{\partial \theta}\right)^2\right]}$$

If equality holds then $\hat{\theta}$ is a minimum variance unbiased estimator of θ.

Efficiency: Let $\hat{\theta}_1$ and $\hat{\theta}_2$ be unbiased estimators of θ.

1. If $\text{Var}(\hat{\theta}_1) < \text{Var}(\hat{\theta}_2)$ then $\hat{\theta}_1$ is relatively more efficient than $\hat{\theta}_2$.

2. The efficiency of $\hat{\theta}_1$ relative to $\hat{\theta}_2$ is

$$\text{Efficiency} = \frac{\text{Var}(\hat{\theta}_2)}{\text{Var}(\hat{\theta}_1)}$$

Consistency: $\hat{\theta}$ is a consistent estimator of θ if for every $\epsilon > 0$,

$$\lim_{n \to \infty} P(|\hat{\theta} - \theta| \leq \epsilon) = 1 \quad \text{or equivalently} \quad \lim_{n \to \infty} P(|\hat{\theta} - \theta| > \epsilon) = 0$$

Theorem: $\hat{\theta}$ is a consistent estimator of θ if

1. $\hat{\theta}$ is unbiased, and

2. $\lim_{n \to \infty} \text{Var}(\hat{\theta}) = 0$.

Sufficiency: $\hat{\theta}$ is a sufficient estimator of θ if for each value of $\hat{\theta}$ the conditional distribution of $X_1, X_2, \ldots, X_n$ given $\hat{\theta}$ equals a specific value is independent of θ.

Theorem: $\hat{\theta}$ is a sufficient estimator of θ if the joint distribution of $X_1, X_2, \ldots, X_n$ can be factored into

$$f(x_1, x_2, \ldots, x_n; \theta) = g(\hat{\theta}, \theta) \cdot h(x_1, x_2, \ldots, x_n)$$

Table 4. Probability and Statistics Formulas (Continued)

where $g(\hat{\theta}, \theta)$ depends only on the estimate $\hat{\theta}$ and the parameter θ, and $h(x_1, x_2, \ldots, x_n)$ does not depend on the parameter θ.

The Method Of Moments: The moment estimators are the solutions to the system of equations

$$\mu'_k = E(X^k) = \frac{1}{n} \sum_{i=1}^{n} x_i^k = m'_k, \quad k = 1, 2, \ldots, r$$

where r is the number of parameters.

The Likelihood Function: Let $x_1, x_2, \ldots, x_n$ be the values of a random sample from a population characterized by the parameters $\theta_1, \theta_2, \ldots, \theta_r$. The likelihood function of the sample is

1. the joint probability mass function evaluated at $x_1, x_2, \ldots, x_n$ if $X_1, X_2, \ldots, X_n$ are discrete,

$$L(\theta_1, \theta_2, \ldots, \theta_r) = p(x_1, x_2, \ldots, x_n; \theta_1, \theta_2, \ldots, \theta_r)$$

2. the joint probability density function evaluated at $x_1, x_2, \ldots, x_n$ if $X_1, X_2, \ldots, X_n$ are continuous.

$$L(\theta_1, \theta_2, \ldots, \theta_r) = f(x_1, x_2, \ldots, x_n; \theta_1, \theta_2, \ldots, \theta_r)$$

The Method Of Maximum Likelihood: The maximum likelihood estimators are those values of the parameters that maximize the likelihood function of the sample $L(\theta_1, \theta_2, \ldots, \theta_r)$.

In practice it is often easier to maximize $\ln L(\theta_1, \theta_2, \ldots, \theta_r)$. This is equivalent to maximizing the likelihood function, $L(\theta_1, \theta_2, \ldots, \theta_r)$, since $\ln L(\theta_1, \theta_2, \ldots, \theta_r)$ is a monotonic function of $L(\theta_1, \theta_2, \ldots, \theta_r)$.

The Invariance Property of Maximum Likelihood Estimators: Let $\hat{\theta}_1, \hat{\theta}_2, \ldots, \hat{\theta}_r$ be the maximum likelihood estimators for $\theta_1, \theta_2, \ldots, \theta_r$ and let $h(\theta_1, \theta_2, \ldots, \theta_r)$ be a function of $\theta_1, \theta_2, \ldots, \theta_r$. The maximum likelihood estimator of the parameter $h(\theta_1, \theta_2, \ldots, \theta_r)$ is $\widehat{h(\theta_1, \theta_2, \ldots, \theta_r)} = h(\hat{\theta}_1, \hat{\theta}_2, \ldots, \hat{\theta}_r)$.

Table 4. Probability and Statistics Formulas (Continued)

Confidence Intervals

Parameter	Assumptions	$100(1-\alpha)\%$ Confidence Interval
μ	n large, σ^2 known, or normality, σ^2 known	$\bar{x} \pm z_{\alpha/2} \cdot \frac{\sigma}{\sqrt{n}}$
μ	n large, σ^2 unknown	$\bar{x} \pm z_{\alpha/2} \cdot \frac{s}{\sqrt{n}}$
μ	normality, n small, σ^2 unknown	$\bar{x} \pm t_{\alpha/2,n-1} \cdot \frac{s}{\sqrt{n}}$
p	binomial experiment, n large	$\hat{p} \pm z_{\alpha/2} \cdot \sqrt{\frac{\hat{p}\hat{q}}{n}}$
σ^2	normality	$\left(\dfrac{(n-1)s^2}{\chi^2_{\alpha/2,n-1}}, \dfrac{(n-1)s^2}{\chi^2_{1-\alpha/2,n-1}} \right)$
$\mu_1 - \mu_2$	n_1, n_2 large, independence, σ_1^2, σ_2^2 known, or normality, independence, σ_1^2, σ_2^2 known	$(\bar{x}_1 - \bar{x}_2) \pm z_{\alpha/2} \cdot \sqrt{\frac{\sigma_1^2}{n_1} + \frac{\sigma_2^2}{n_2}}$
$\mu_1 - \mu_2$	n_1, n_2 large, independence, σ_1^2, σ_2^2 unknown	$(\bar{x}_1 - \bar{x}_2) \pm z_{\alpha/2} \cdot \sqrt{\frac{s_1^2}{n_1} + \frac{s_2^2}{n_2}}$
$\mu_1 - \mu_2$	normality, independence, σ_1^2, σ_2^2 unknown but equal, n_1, n_2 small	$(\bar{x}_1 - \bar{x}_2) \pm t_{\alpha/2,n_1+n_2-2} \cdot s_p \sqrt{\frac{1}{n_1} + \frac{1}{n_2}}$ $s_p = \dfrac{(n_1-1)s_1^2 + (n_2-1)s_2^2}{n_1 + n_2 - 2}$
$\mu_1 - \mu_2$	normality, independence, σ_1^2, σ_2^2 unknown, unequal, n_1, n_2 small	$(\bar{x}_1 - \bar{x}_2) \pm t_{\alpha/2,\nu} \cdot \sqrt{\frac{s_1^2}{n_1} + \frac{s_2^2}{n_2}}$ $\nu = \dfrac{\left(\frac{s_1^2}{n_1} + \frac{s_2^2}{n_2} \right)^2}{\frac{(s_1^2/n_1)^2}{n_1-1} + \frac{(s_2^2/n_2)^2}{n_2-1}}$
$\mu_D = \mu_1 - \mu_2$	normality, n pairs, n small, dependence	$\bar{d} \pm t_{\alpha/2,n-1} \cdot \frac{s_D}{\sqrt{n}}$
$p_1 - p_2$	binomial experiments, n_1, n_2 large, independence	$(\hat{p}_1 - \hat{p}_2) \pm z_{\alpha/2} \cdot \sqrt{\frac{\hat{p}_1\hat{q}_1}{n_1} + \frac{\hat{p}_2\hat{q}_2}{n_2}}$
$\dfrac{\sigma_1^2}{\sigma_2^2}$	normality, independence	$\left(\dfrac{s_1^2}{s_2^2} \cdot \dfrac{1}{F_{\frac{\alpha}{2},n_1-1,n_2-1}}, \dfrac{s_1^2}{s_2^2} \cdot \dfrac{1}{F_{1-\frac{\alpha}{2},n_1-1,n_2-1}} \right)$

Table 4. Probability and Statistics Formulas (Continued)

Hypothesis Tests (One-Sample)

Null Hypothesis	Assumptions	Alternative Hypothesis	Test Statistic	Rejection Region
$\mu = \mu_0$	n large, σ^2 known, or normality, σ^2 known	$\mu > \mu_0$ $\mu < \mu_0$ $\mu \neq \mu_0$	$Z = \dfrac{\overline{X} - \mu_0}{\sigma/\sqrt{n}}$	$Z \geq z_\alpha$ $Z \leq -z_\alpha$ $\mid Z \mid \geq z_{\alpha/2}$
$\mu = \mu_0$	n large, σ^2 unknown	$\mu > \mu_0$ $\mu < \mu_0$ $\mu \neq \mu_0$	$Z = \dfrac{\overline{X} - \mu_0}{s/\sqrt{n}}$	$Z \geq z_\alpha$ $Z \leq -z_\alpha$ $\mid Z \mid \geq z_{\alpha/2}$
$\mu = \mu_0$	normality, n small, σ^2 unknown	$\mu > \mu_0$ $\mu < \mu_0$ $\mu \neq \mu_0$	$T = \dfrac{\overline{X} - \mu_0}{S/\sqrt{n}}$	$T \geq t_{\alpha,n-1}$ $T \leq -t_{\alpha,n-1}$ $\mid T \mid \geq t_{\alpha/2,n-1}$
$p = p_0$	binomial experiment, n large	$p > p_0$ $p < p_0$ $p \neq p_0$	$Z = \dfrac{\hat{p} - p_0}{\sqrt{p_0(1-p_0)/n}}$	$Z \geq z_\alpha$ $Z \leq -z_\alpha$ $\mid Z \mid \geq z_{\alpha/2}$
$\sigma^2 = \sigma_0^2$	normality	$\sigma^2 > \sigma_0^2$ $\sigma^2 < \sigma_0^2$ $\sigma^2 \neq \sigma_0^2$	$\chi^2 = \dfrac{(n-1)S^2}{\sigma_0^2}$	$\chi^2 \geq \chi^2_{\alpha,n-1}$ $\chi^2 \leq \chi^2_{1-\alpha,n-1}$ $\chi^2 \leq \chi^2_{1-\alpha/2,n-1}$ or $\chi^2 \geq \chi^2_{\alpha/2,n-1}$

Table 4. Probability and Statistics Formulas (Continued)

Hypothesis Tests (Two-Samples)

Null Hypothesis	Assumptions	Alternative Hypothesis	Test Statistic	Rejection Region
$\mu_1 - \mu_2 = \Delta_0$	n_1, n_2 large, independence, σ_1^2, σ_2^2 known, or normality, independence, σ_1^2, σ_2^2 known	$\mu_1 - \mu_2 > \Delta_0$ $\mu_1 - \mu_2 < \Delta_0$ $\mu_1 - \mu_2 \neq \Delta_0$	$Z = \dfrac{(\overline{X}_1 - \overline{X}_2) - \Delta_0}{\sqrt{\frac{\sigma_1^2}{n_1} + \frac{\sigma_2^2}{n_2}}}$	$Z \geq z_\alpha$ $Z \leq -z_\alpha$ $\lvert Z \rvert \geq z_{\alpha/2}$
$\mu_1 - \mu_2 = \Delta_0$	n_1, n_2 large, independence, σ_1^2, σ_2^2 unknown	$\mu_1 - \mu_2 > \Delta_0$ $\mu_1 - \mu_2 < \Delta_0$ $\mu_1 - \mu_2 \neq \Delta_0$	$Z = \dfrac{(\overline{X}_1 - \overline{X}_2) - \Delta_0}{\sqrt{\frac{s_1^2}{n_1} + \frac{s_2^2}{n_2}}}$	$Z \geq z_\alpha$ $Z \leq -z_\alpha$ $\lvert Z \rvert \geq z_{\alpha/2}$
$\mu_1 - \mu_2 = \Delta_0$	normality, independence, σ_1^2, σ_2^2 unknown, $\sigma_1^2 = \sigma_2^2$ n_1, n_2 small	$\mu_1 - \mu_2 > \Delta_0$ $\mu_1 - \mu_2 < \Delta_0$ $\mu_1 - \mu_2 \neq \Delta_0$	$T = \dfrac{(\overline{X}_1 - \overline{X}_2) - \Delta_0}{S_p \sqrt{\frac{1}{n_1} + \frac{1}{n_2}}}$ $S_p = \dfrac{(n_1 - 1)S_1^2 + (n_2 - 1)S_2^2}{n_1 + n_2 - 2}$	$T \geq t_{\alpha, n_1 + n_2 - 2}$ $T \leq -t_{\alpha, n_1 + n_2 - 2}$ $\lvert T \rvert \geq t_{\alpha/2, n_1 + n_2 - 2}$
$\mu_1 - \mu_2 = \Delta_0$	normality, independence, σ_1^2, σ_2^2 unknown, $\sigma_1^2 \neq \sigma_2^2$ n_1, n_2 small	$\mu_1 - \mu_2 > \Delta_0$ $\mu_1 - \mu_2 < \Delta_0$ $\mu_1 - \mu_2 \neq \Delta_0$	$T' = \dfrac{(\overline{X}_1 - \overline{X}_2) - \Delta_0}{\sqrt{\frac{S_1^2}{n_1} + \frac{S_2^2}{n_2}}}$ $\nu = \dfrac{\left(\frac{s_1^2}{n_1} + \frac{s_2^2}{n_2}\right)^2}{\frac{(s_1^2/n_1)^2}{n_1-1} + \frac{(s_2^2/n_2)^2}{n_2-1}}$	$T' \geq t_{\alpha/2, \nu}$ $T' \leq -t_{\alpha/2, \nu}$ $\lvert T' \rvert \geq t_{\alpha/2, \nu}$
$\mu_D = \Delta_0$	normality, n pairs, n small, dependence	$\mu_D > \Delta_0$ $\mu_D < \Delta_0$ $\mu_D \neq \Delta_0$	$T = \dfrac{\overline{D} - \Delta_0}{S_D / \sqrt{n}}$	$T \geq t_{\alpha, n-1}$ $T \leq -t_{\alpha, n-1}$ $\lvert T \rvert \geq t_{\alpha/2, n-1}$
$p_1 - p_2 = 0$	binomial exps., n_1, n_2 large, independence	$p_1 - p_2 > 0$ $p_1 - p_2 < 0$ $p_1 - p_2 \neq 0$	$Z = \dfrac{\hat{p}_1 - \hat{p}_2}{\sqrt{\hat{p}\hat{q}(1/n_1 + 1/n_2)}}$ $\hat{p} = \dfrac{X_1 + X_2}{n_1 + n_2}$	$Z \geq z_\alpha$ $Z \leq -z_\alpha$ $\lvert Z \rvert \geq z_{\alpha/2}$
$p_1 - p_2 = \Delta_0$	binomial exps., n_1, n_2 large, independence	$p_1 - p_2 > \Delta_0$ $p_1 - p_2 < \Delta_0$ $p_1 - p_2 \neq \Delta_0$	$Z = \dfrac{(\hat{p}_1 - \hat{p}_2) - \Delta_0}{\sqrt{\frac{\hat{p}_1\hat{q}_1}{n_1} + \frac{\hat{p}_2\hat{q}_2}{n_2}}}$	$Z \geq z_\alpha$ $Z \leq -z_\alpha$ $\lvert Z \rvert \geq z_{\alpha/2}$
$\sigma_1^2 = \sigma_2^2$	normality, independence	$\sigma_1^2 > \sigma_2^2$ $\sigma_1^2 < \sigma_2^2$ $\sigma_1^2 \neq \sigma_2^2$	$F = S_1^2 / S_2^2$	$F \geq F_{\alpha, n_1-1, n_2-1}$ $F \leq F_{1-\alpha, n_1-1, n_2-1}$ $F \leq F_{1-\frac{\alpha}{2}, n_1-1, n_2-1}$ or $F \geq F_{\frac{\alpha}{2}, n_1-1, n_2-1}$

Table 4. Probability and Statistics Formulas (Continued)

Hypothesis Tests

Type I Error: Rejecting the null hypothesis when it is true is a type I error.

$$\alpha = P(\text{type I error}) = \text{Significance level} = P(\text{rejecting} H_0 \mid H_0 \text{ is true})$$

Type II Error: Accepting the null hypothesis when it is false is a type II error.

$$\beta = P(\text{type II error}) = P(\text{accepting} H_0 \mid H_0 \text{ is false})$$

The Power Function: The power function of a statistical test of H_0 versus the alternative H_a is

$$\pi(\theta) = \begin{cases} \alpha(\theta) & \text{for values of } \theta \text{ assumed under } H_0 \\ 1 - \beta(\theta) & \text{for values of } \theta \text{ assumed under } H_a \end{cases}$$

The p-Value: The p-value of a statistical test is the smallest α level for which H_0 can be rejected.

The Neyman-Pearson Lemma: Given the null hypothesis $H_0 : \theta = \theta_0$ versus the alternative hypothesis $H_a : \theta = \theta_a$, let $L(\theta)$ be the likelihood function evaluated at θ. For a given α, the test that maximizes the power at θ_a has a rejection region determined by

$$\frac{L(\theta_0)}{L(\theta_a)} < k$$

This statistical test is the most powerful test of H_0 versus H_a.

Likelihood Ratio Tests: Given the null hypothesis $H_0 : \underline{\theta} \in \Omega_0$ versus the alternative hypothesis $H_a : \underline{\theta} \in \Omega_a$, $\Omega_0 \cap \Omega_a = \emptyset$, $\Omega = \Omega_0 \cup \Omega_a$. Let $L(\hat{\Omega}_0)$ be the likelihood function with all unknown parameters replaced by their maximum likelihood estimators subject to the constraint $\underline{\theta} \in \Omega_0$, and let $L(\hat{\Omega})$ be defined similarly subject to the constraint $\underline{\theta} \in \Omega$. Define

$$\lambda = \frac{L(\hat{\Omega}_0)}{L(\hat{\Omega})}$$

A likelihood ratio test of H_0 versus H_a uses λ as a test statistic and has a rejection region given by $\lambda \leq k$, $0 < k < 1$.

Under very general conditions, for large n, $-2 \ln \lambda$ has approximately a chi-square distribution with degrees of freedom equal to the number of parameters or functions of parameters with specific values under H_0.

Goodness of Fit Test: Let n_i be the number of observations falling into the ith category, $i = 1, 2, \ldots, k$, and let $n = n_1 + n_2 + \cdots + n_k$.

$H_0 : p_1 = p_{10}, p_2 = p_{20}, \ldots, p_k = p_{k0}$

$H_a : p_i \neq p_{i0}$ for at least one i

Test Statistic: $\chi^2 = \displaystyle\sum_{i=1}^{k} \frac{(\text{observed} - \text{estimated expected})^2}{\text{estimated expected}} = \sum_{i=1}^{k} \frac{(n_i - np_{i0})^2}{np_{i0}}$

Under the null hypothesis χ^2 has approximately a chi-square distribution with $k - 1$ degrees of freedom. The approximation is satisfactory if $np_{i0} \geq 5$ for all i.

Rejection Region: $\chi^2 \geq \chi^2_{\alpha, k-1}$

Contingency Tables: Let the contingency table contain I rows and J columns, let n_{ij} be the count in the (i, j)th cell, and let $\hat{e}_{ij}$ be the estimated expected count in that cell. The test statistic is

$$\chi^2 = \sum_{\text{all cells}} \frac{(\text{observed} - \text{estimated expected})^2}{\text{estimated expected}} = \sum_{i=1}^{I} \sum_{j=1}^{J} \frac{(n_{ij} - \hat{e}_{ij})^2}{\hat{e}_{ij}}$$

Table 4. Probability and Statistics Formulas (Continued)

where $\hat{e}_{ij} = \dfrac{(i\text{th row total})(j\text{th column total})}{\text{grand total}} = \dfrac{n_{i.}\,n_{.j}}{n}$

Under the null hypothesis χ^2 has approximately a chi-square distribution with $(I-1)(J-1)$ degrees of freedom. The approximation is satisfactory if $\hat{e}_{ij} \geq 5$ for all i and j.

Bartlett's Test: Let there be k independent samples with n_i, $i = 1, 2, \ldots, k$ observations in each sample, $N = n_1 + n_2 + \cdots + n_k$, and let S_i^2 be the ith sample variance.

$H_0 : \sigma_1^2 = \sigma_2^2 = \cdots = \sigma_k^2$

H_a : the variances are not all equal

Test Statistic: $B = \dfrac{[(S_1^2)^{n_1-1}(S_2^2)^{n_2-1}\cdots(S_k^2)^{n_k-1}]^{1/(N-k)}}{S_p^2}$ where $S_p^2 = \dfrac{\sum\limits_{i=1}^{k}(n_i - 1)S_i^2}{N - k}$

Rejection Region $(n_1 = n_2 = \cdots = n_k = n)$: $B \leq b_{\alpha,k,n}$

Rejection Region (sample sizes unequal): $B \leq b_{\alpha,k,n_1,n_2,\ldots,n_k}$

where $b_{\alpha,k,n_1,n_2,\ldots,n_k} \approx \dfrac{n_1 b_{\alpha,k,n_1} + n_2 b_{\alpha,k,n_2} + \cdots + n_k b_{\alpha,k,n_k}}{N}$

Approximate Test Procedure: Let $\nu_i = n_i - 1$

Test Statistic: $\chi^2 = M/C$ where

$$M = \left(\sum_{i=1}^{k}\nu_i\right)\ln \overline{S}^2 - \sum_{i=1}^{k}\ln S_i^2 \quad\text{and}\quad \overline{S}^2 = \sum_{i=1}^{k}\nu_i S_i^2 \Big/ \sum_{i=1}^{k}\nu_i$$

$$C = 1 + \frac{1}{3(k-1)}\left(\sum_{i=1}^{k}1/\nu_i - 1\Big/\sum_{i=1}^{k}\nu_i\right)$$

Under the null hypothesis χ^2 has approximately a chi-square distribution with $k-1$ degrees of freedom.

Rejection Region: $\chi^2 \geq \chi^2_{\alpha,k-1}$

Cochran's Test: Let there be k independent samples with n observations in each sample, and let S_i^2 be the ith sample variance, $i = 1, 2, \ldots, k$.

$H_0 : \sigma_1^2 = \sigma_2^2 = \cdots = \sigma_k^2$

H_a : the variances are not all equal

Test Statistic: $G = \dfrac{\text{largest } S_i^2}{\sum\limits_{i=1}^{k} S_i^2}$

Rejection Region: $G \geq g_{\alpha,k,n}$

Simple Linear Regression

The Model: Let $(x_1, y_1), (x_2, y_2), \ldots, (x_n, y_n)$ be n pairs of observations such that y_i is an observed value of the random variable Y_i. We assume there exist constants β_0 and β_1 such that

$$Y_i = \beta_0 + \beta_1 x_i + \epsilon_i$$

where $\epsilon_1, \epsilon_2, \ldots, \epsilon_n$ are independent, normal random variables having mean 0 and variance σ^2. That is

1. The ϵ_i's are normally distributed   (the Y_i's are normally distributed),
2. $E(\epsilon_i) = 0$ $(E(Y_i) = \beta_0 + \beta_1 x_i)$,

Table 4. Probability and Statistics Formulas (Continued)

3. $\text{Var}(\epsilon_i) = \sigma^2$ $(\text{Var}(Y_i) = \sigma^2)$, and
4. $\text{Cov}(\epsilon_i, \epsilon_j) = 0,\ i \neq j$ $(\text{Cov}(Y_i, Y_j) = 0,\ i \neq j)$.

Principle Of Least Squares: The sum of squared deviations about the true regression line is

$$S(\beta_0, \beta_1) = \sum_{i=1}^{n}[y_i - (\beta_0 + \beta_1 x_i)]^2$$

The point estimates of β_0 and β_1, denoted by $\hat{\beta}_0$ and $\hat{\beta}_1$, are those values that minimize $S(\beta_0, \beta_1)$. $\hat{\beta}_0$ and $\hat{\beta}_1$ are called the least squares estimates. The estimated regression line or least squares line is $y = \hat{\beta}_0 + \hat{\beta}_1 x$.

Normal Equations:

$$\sum_{i=1}^{n} y_i = n\hat{\beta}_0 + \hat{\beta}_1 \sum_{i=1}^{n} x_i$$

$$\sum_{i=1}^{n} x_i y_i = \hat{\beta}_0 \sum_{i=1}^{n} x_i + \hat{\beta}_1 \sum_{i=1}^{n} x_i^2$$

Notation:

$$S_{xx} = \sum_{i=1}^{n}(x_i - \bar{x})^2 = \sum_{i=1}^{n} x_i^2 - \frac{\left(\sum_{i=1}^{n} x_i\right)^2}{n}$$

$$S_{yy} = \sum_{i=1}^{n}(y_i - \bar{y})^2 = \sum_{i=1}^{n} y_i^2 - \frac{\left(\sum_{i=1}^{n} y_i\right)^2}{n}$$

$$S_{xy} = \sum_{i=1}^{n}(x_i - \bar{x})(y_i - \bar{y}) = \sum_{i=1}^{n} x_i y_i - \frac{\left(\sum_{i=1}^{n} x_i\right)\left(\sum_{i=1}^{n} y_i\right)}{n}$$

Least Squares Estimates:

$$\hat{\beta}_1 = \frac{S_{xy}}{S_{xx}} = \frac{n\sum_{i=1}^{n} x_i y_i - \left(\sum_{i=1}^{n} x_i\right)\left(\sum_{i=1}^{n} y_i\right)}{n\sum_{i=1}^{n} x_i^2 - \left(\sum_{i=1}^{n} x_i\right)^2} \qquad \hat{\beta}_0 = \frac{\sum_{i=1}^{n} y_i - \hat{\beta}_1 \sum_{i=1}^{n} x_i}{n} = \bar{y} - \hat{\beta}_1 \bar{x}$$

The *ith predicted (fitted) value*: $\hat{y}_i = \hat{\beta}_0 + \hat{\beta}_1 x_i,\ i = 1, 2, \ldots, n$

The *ith residual*: $e_i = y_i - \hat{y}_i,\ i = 1, 2, \ldots, n$

Properties:

1. $E(\hat{\beta}_1) = \beta_1,\quad \text{Var}(\hat{\beta}_1) = \dfrac{\sigma^2}{\sum_{i=1}^{n}(x_i - \bar{x})^2} = \dfrac{\sigma^2}{S_{xx}}$

2. $E(\hat{\beta}_0) = \beta_0,\quad \text{Var}(\hat{\beta}_0) = \dfrac{\sigma^2 \sum_{i=1}^{n} x_i}{n\sum_{i=1}^{n}(x_i - \bar{x})^2} = \dfrac{\sigma^2 \sum_{i=1}^{n} x_i}{nS_{xx}}$

3. $\hat{\beta}_0$ and $\hat{\beta}_1$ are normally distributed.

Table 4. Probability and Statistics Formulas (Continued)

The Sum Of Squares:

$$\sum_{i=1}^{n}(y_i - \bar{y})^2 = \sum_{i=1}^{n}(\hat{y}_i - \bar{y})^2 + \sum_{i=1}^{n}(y_i - \hat{y}_i)^2$$

$$\underbrace{\phantom{\sum_{i=1}^{n}(y_i - \bar{y})^2}}_{SST} \quad \underbrace{\phantom{\sum_{i=1}^{n}(\hat{y}_i - \bar{y})^2}}_{SSR} \quad \underbrace{\phantom{\sum_{i=1}^{n}(y_i - \hat{y}_i)^2}}_{SSE}$$

SST = total sum of squares = S_{yy}

SSR = sum of squares due to regression = $\hat{\beta}_1 S_{xy}$

SSE = sum of squares due to error

$$= \sum_{i=1}^{n}[y_i - (\hat{\beta}_0 + \hat{\beta}_1 x_i)]^2 = \sum_{i=1}^{n}y_i^2 - \hat{\beta}_0 \sum_{i=1}^{n}y_i - \hat{\beta}_1 \sum_{i=1}^{n}x_i y_i$$

$$= S_{yy} - 2\hat{\beta}_1 S_{xy} + \hat{\beta}_1^2 S_{xx} = S_{yy} - \hat{\beta}_1^2 S_{xx} = S_{yy} - \hat{\beta}_1 S_{xy}$$

1. $\hat{\sigma}^2 = s^2 = \dfrac{SSE}{n-2}, \qquad E(S^2) = \sigma^2$

2. *Sample Coefficient of Determination:* $r^2 = \dfrac{SSR}{SST} = 1 - \dfrac{SSE}{SST}$

Inferences Concerning The Regression Coefficients:

The Parameter β_1:

1. $T = \dfrac{\hat{\beta}_1 - \beta_1}{S/\sqrt{S_{xx}}} = \dfrac{\hat{\beta}_1 - \beta_1}{S_{\hat{\beta}_1}}$ has a t distribution with $n-2$ degrees of freedom.

2. A $100(1-\alpha)\%$ confidence interval for β_1 has as endpoints $\hat{\beta}_1 \pm t_{\alpha/2, n-2} \cdot s_{\hat{\beta}_1}$

3. Hypothesis test

Null Hypothesis	Alternative Hypothesis	Test Statistic	Rejection Region
$\beta_1 = \beta_{10}$	$\beta_1 > \beta_{10}$ $\beta_1 < \beta_{10}$ $\beta_1 \neq \beta_{10}$	$T = \dfrac{\hat{\beta}_1 - \beta_{10}}{S_{\hat{\beta}_1}}$	$T \geq t_{\alpha, n-2}$ $T \leq -t_{\alpha, n-2}$ $\mid T \mid \geq t_{\alpha/2, n-2}$

The Parameter β_0:

1. $T = \dfrac{\hat{\beta}_0 - \beta_0}{S\sqrt{\sum_{i=1}^{n} x_i^2 / nS_{xx}}} = \dfrac{\hat{\beta}_0 - \beta_0}{S_{\hat{\beta}_0}}$ has a t distribution with $n-2$ degrees of freedom.

2. A $100(1-\alpha)\%$ confidence interval for β_0 has as endpoints $\hat{\beta}_0 \pm t_{\alpha/2, n-2} \cdot s_{\hat{\beta}_0}$

3. Hypothesis test

Null Hypothesis	Alternative Hypothesis	Test Statistic	Rejection Region
$\beta_0 = \beta_{00}$	$\beta_0 > \beta_{00}$ $\beta_0 < \beta_{00}$ $\beta_0 \neq \beta_{00}$	$T = \dfrac{\hat{\beta}_0 - \beta_{00}}{S_{\hat{\beta}_0}}$	$T \geq t_{\alpha, n-2}$ $T \leq -t_{\alpha, n-2}$ $\mid T \mid \geq t_{\alpha/2, n-2}$

The Mean Response: The mean response of Y given $x = x_0$ is $\mu_{Y|x_0} = \beta_0 + \beta_1 x_0$. The random variable $\hat{Y}_0 = \hat{\beta}_0 + \hat{\beta}_1 x_0$ is used to estimate $\mu_{Y|x_0}$.

Table 4. Probability and Statistics Formulas (Continued)

1. $E(\hat{Y}_0) = \beta_0 + \beta_1 x_0$

2. $\mathrm{Var}(\hat{Y}_0) = \sigma^2 \left[\dfrac{1}{n} + \dfrac{(x_0 - \bar{x})^2}{S_{xx}} \right]$

3. $\hat{Y}_0$ has a normal distribution.

4. $T = \dfrac{\hat{Y}_0 - \mu_{Y|x_0}}{S\sqrt{(1/n) + [(x_0 - \bar{x})^2/S_{xx}]}} = \dfrac{\hat{Y}_0 - \mu_{Y|x_0}}{S_{\hat{Y}_0}}$ has a t distribution with $n - 2$ degrees of freedom.

5. A $100(1 - \alpha)\%$ confidence interval for $\mu_{Y|x_0}$ has as endpoints $\hat{y}_0 \pm t_{\alpha/2, n-2} \cdot s_{\hat{Y}_0}$.

6. Hypothesis test

Null Hypothesis	Alternative Hypothesis	Test Statistic	Rejection Region
$\beta_0 + \beta_1 x_0 = y_0 = \mu_0$	$y_0 > \mu_0$		$T \geq t_{\alpha, n-2}$
	$y_0 < \mu_0$	$T = \dfrac{\hat{Y}_0 - \mu_0}{S_{\hat{Y}_0}}$	$T \leq -t_{\alpha, n-2}$
	$y_0 \neq \mu_0$		$|T| \geq t_{\alpha/2, n-2}$

Prediction Interval: A prediction interval for a value y_0 of the random variable $Y_0 = \beta_0 + \beta_1 x_0 + \epsilon_0$ is obtained by considering the random variable $\hat{Y}_0 - Y_0$.

1. $E(\hat{Y}_0 - Y_0) = 0$

2. $\mathrm{Var}(\hat{Y}_0 - Y_0) = \sigma^2 \left[1 + \dfrac{1}{n} + \dfrac{(x_0 - \bar{x})^2}{S_{xx}} \right]$

3. $\hat{Y}_0 - Y_0$ has a normal distribution.

4. $T = \dfrac{\hat{Y}_0 - Y_0}{S\sqrt{1 + (1/n) + [(x_0 - \bar{x})^2/S_{xx}]}} = \dfrac{\hat{Y}_0 - Y_0}{S_{\hat{Y}_0 - Y_0}}$ has a t distribution with $n - 2$ degrees of freedom.

5. A $100(1 - \alpha)\%$ prediction interval for y_0 has as endpoints $\hat{y}_0 \pm t_{\alpha/2, n-2} \cdot s_{\hat{Y}_0 - Y_0}$

Analysis Of Variance Table:

Source of Variation	Sum of Squares	Degrees of Freedom	Mean Square	Computed F
Regression	SSR	1	$MSR = \frac{SSR}{1}$	MSR/MSE
Error	SSE	$n - 2$	$MSE = \frac{SSE}{n-2}$	
Total	SST	$n - 1$		

Hypothesis Test of Significant Regression:

Null Hypothesis	Alternative Hypothesis	Test Statistic	Rejection Region
$\beta_1 = 0$	$\beta_1 \neq 0$	$F = MSR/MSE$	$F \geq F_{\alpha, 1, n-2}$

Test For Linearity Of Regression: Let there be k distinct values of x, $\{x_1, x_2, \ldots, x_k\}$, n_i observations for x_i, and $n = n_1 + n_2 + \cdots + n_k$. Define

$$y_{ij} = \text{the } j\text{th observation on the random variable } Y_i, \quad T_i = \sum_{j=1}^{n_i} y_{ij}, \quad \bar{y}_{i.} = T_i / n_i$$

Table 4. Probability and Statistics Formulas (Continued)

$$SSPE = \text{sum of squares due to pure error} = \sum_{i=1}^{k}\sum_{j=1}^{n_i}(y_{ij} - \bar{y}_{i.})^2 = \sum_{i=1}^{k}\sum_{j=1}^{n_i}y_{ij}^2 - \sum_{i=1}^{k}\frac{T_i^2}{n_i}$$

$SSLF = \text{sum of squares due to lack of fit} = SSE - SSPE$

Test Statistic: $F = \dfrac{SSLF/(k-2)}{SSPE/(n-k)}$

Rejection Region: $F \geq F_{\alpha,k-2,n-k}$

Sample Correlation Coefficient: $r = \hat{\beta}_1\sqrt{\dfrac{S_{xx}}{S_{yy}}} = \dfrac{S_{xy}}{\sqrt{S_{xx}S_{yy}}}$

Hypothesis tests

Null Hypothesis	Alternative Hypothesis	Test Statistic	Rejection Region
$\rho = 0$	$\rho > 0$		$T \geq t_{\alpha,n-2}$
	$\rho < 0$	$T = \dfrac{R\sqrt{n-2}}{\sqrt{1-R^2}} = \hat{\beta}_1/S_{\hat{\beta}_1}$	$T \leq -t_{\alpha,n-2}$
	$\rho \neq 0$		$\lvert T \rvert \geq t_{\alpha/2,n-2}$

If X and Y have a bivariate normal distribution:

$\rho = \rho_0$	$\rho > \rho_0$		$Z \geq z_\alpha$
	$\rho < \rho_0$	$Z = \dfrac{\sqrt{n-3}}{2}\ln\left[\dfrac{(1+R)(1-\rho_0)}{(1-R)(1+\rho_0)}\right]$	$Z \leq -z_\alpha$
	$\rho \neq \rho_0$		$\lvert Z \rvert \geq z_{\alpha/2}$

Multiple Linear Regression

The Model: Let there be n observations of the form $(x_{1i}, x_{2i}, \ldots, x_{ki}, y_i)$ such that y_i is an observed value of the random variable Y_i. Assume there exist constants $\beta_0, \beta_1, \ldots, \beta_k$ such that

$$Y_i = \beta_0 + \beta_1 x_{1i} + \cdots + \beta_k x_{ki} + \epsilon_i$$

where $\epsilon_1, \epsilon_2, \ldots, \epsilon_n$ are independent, normal random variables having mean 0 and variance σ^2. That is

1. The ϵ_i's are normally distributed   (the Y_i's are normally distributed),
2. $E(\epsilon_i) = 0$ $(E(Y_i) = \beta_0 + \beta_1 x_{1i} + \cdots + \beta_k x_{ki})$,
3. $\text{Var}(\epsilon_i) = \sigma^2$, $(\text{Var}(Y_i) = \sigma^2)$, and
4. $\text{Cov}(\epsilon_i, \epsilon_j) = 0$, $i \neq j$, $(\text{Cov}(Y_i, Y_j) = 0,\ i \neq j)$.

Notation: Let $\mathbf{Y}$ be the random vector of responses, $\mathbf{y}$ be the vector of observed responses, $\boldsymbol{\beta}$ be the vector of regression coefficients, $\boldsymbol{\epsilon}$ be the vector of random errors, and let $\mathbf{X}$ be the design matrix:

$$\mathbf{Y} = \begin{pmatrix} Y_1 \\ Y_2 \\ \vdots \\ Y_n \end{pmatrix} \quad \mathbf{y} = \begin{pmatrix} y_1 \\ y_2 \\ \vdots \\ y_n \end{pmatrix} \quad \boldsymbol{\beta} = \begin{pmatrix} \beta_0 \\ \beta_1 \\ \vdots \\ \beta_k \end{pmatrix} \quad \boldsymbol{\epsilon} = \begin{pmatrix} \epsilon_1 \\ \epsilon_2 \\ \vdots \\ \epsilon_n \end{pmatrix} \quad \mathbf{X} = \begin{pmatrix} 1 & x_{11} & x_{21} & \cdots & x_{k1} \\ 1 & x_{12} & x_{22} & \cdots & x_{k2} \\ \vdots & \vdots & \vdots & & \vdots \\ 1 & x_{1n} & x_{2n} & \cdots & x_{kn} \end{pmatrix}$$

The model can now be written as: $\mathbf{Y} = \mathbf{X}\boldsymbol{\beta} + \boldsymbol{\epsilon}$

where $\boldsymbol{\epsilon} \sim N_n(\mathbf{0}, \sigma^2\mathbf{I}_n)$ equivalently, $\mathbf{Y} \sim N_n(\mathbf{X}\boldsymbol{\beta}, \sigma^2\mathbf{I}_n)$

Principle Of Least Squares: The sum of squared deviations about the true regression line is

$$S(\boldsymbol{\beta}) = \sum_{i=1}^{n}[y_i - (\beta_0 + \beta_1 x_{1i} + \cdots + \beta_k x_{ki})]^2 = \|\mathbf{y} - \mathbf{X}\boldsymbol{\beta}\|^2$$

The vector $\hat{\boldsymbol{\beta}}' = (\hat{\beta}_0, \hat{\beta}_1, \ldots, \hat{\beta}_k)$ that minimizes $S(\boldsymbol{\beta})$ is the vector of least squares estimates. The estimated regression line or least squares line is $y = \hat{\beta}_0 + \hat{\beta}_1 x_1 + \cdots + \hat{\beta}_k x_k$.

Table 4. Probability and Statistics Formulas (Continued)

Normal Equations: $(\mathbf{X'X})\hat{\boldsymbol{\beta}} = \mathbf{X'y}$

Least Squares Estimates: If the matrix $\mathbf{X'X}$ is non-singular, then $\hat{\boldsymbol{\beta}} = (\mathbf{X'X})^{-1}\mathbf{X'y}$

> The *i*th predicted (fitted) value: $\hat{y}_i = \hat{\beta}_0 + \hat{\beta}_1 x_{1i} + \cdots + \hat{\beta}_k x_{ki}, \quad i = 1, 2, \ldots, n, \quad \hat{\mathbf{y}} = \mathbf{X}\hat{\boldsymbol{\beta}}$

> The *i*th residual: $e_i = y_i - \hat{y}_i, \quad i = 1, 2, \ldots, n, \quad \boldsymbol{\epsilon} = \mathbf{y} - \hat{\mathbf{y}}$

Properties: For $i = 0, 1, 2, \ldots, k, \quad j = 0, 1, 2, \ldots, k$

1. $E(\hat{\beta}_i) = \beta_i$

2. $\text{Var}(\hat{\beta}_i) = c_{ii}\sigma^2$, where c_{ij} is the value in the *i*th row and *j*th column of the matrix $(\mathbf{X'X})^{-1}$.

3. $\hat{\beta}_i$ is normally distributed.

4. $\text{Cov}(\hat{\beta}_i, \hat{\beta}_j) = c_{ij}\sigma^2, \quad i \neq j$

The Sum Of Squares:

$$\sum_{i=1}^{n}(y_i - \bar{y})^2 = \underbrace{\sum_{i=1}^{n}(\hat{y}_i - \bar{y})^2}_{SSR} + \underbrace{\sum_{i=1}^{n}(y_i - \hat{y}_i)^2}_{SSE}$$
$$\underbrace{\phantom{\sum_{i=1}^{n}(y_i - \bar{y})^2}}_{SST}$$

SST = total sum of squares = $\|\mathbf{y} - \bar{y}\mathbf{1}\|^2 = \mathbf{y'y} - n\bar{y}^2$

SSR = sum of squares due to regression = $\|\mathbf{X}\hat{\boldsymbol{\beta}} - \bar{y}\mathbf{1}\|^2 = \hat{\boldsymbol{\beta}}'\mathbf{X'y} - n\bar{y}^2$

SSE = sum of squares due to error = $\|\mathbf{y} - \mathbf{X}\hat{\boldsymbol{\beta}}\|^2 = \mathbf{y'y} - \hat{\boldsymbol{\beta}}'\mathbf{X'y}$

> where $\mathbf{1'} = \underbrace{(1, 1, \ldots, 1)}_{n \, 1's}$

1. $\hat{\sigma}^2 = s^2 = \dfrac{SSE}{n - k - 1}, \qquad E(S^2) = \sigma^2$

2. $\dfrac{(n - k - 1)S^2}{\sigma^2}$ has a chi-square distribution with $n - k - 1$ degrees of freedom, and S^2 and $\hat{\beta}_i$ are independent.

3. *The Coefficient of Multiple Determination:* $R^2 = \dfrac{SSR}{SST} = 1 - \dfrac{SSE}{SST}$

4. *Adjusted Coefficient of Multiple Determination:* $R_a^2 = 1 - \left(\dfrac{n-1}{n-k-1}\right)\dfrac{SSE}{SST} = 1 - (1 - R^2)\left(\dfrac{n-1}{n-k-1}\right)$

Inferences Concerning The Regression Coefficients:

1. $T = \dfrac{\hat{\beta}_i - \beta_i}{S\sqrt{c_{ii}}}$ has a t distribution with $n - k - 1$ degrees of freedom.

2. A $100(1 - \alpha)\%$ confidence for β_i has as endpoints $\hat{\beta}_i \pm t_{\alpha/2, n-k-1} \cdot s\sqrt{c_{ii}}$

3. Hypothesis test for β_i

Null Hypothesis	Alternative Hypothesis	Test Statistic	Rejection Region
$\beta_i = \beta_{i0}$	$\beta_i > \beta_{i0}$	$T = \dfrac{\hat{\beta}_i - \beta_i}{S\sqrt{c_{ii}}}$	$T \geq t_{\alpha, n-k-1}$
	$\beta_i < \beta_{i0}$		$T \leq -t_{\alpha, n-k-1}$
	$\beta_i \neq \beta_{i0}$		$\lvert T \rvert \geq t_{\alpha/2, n-k-1}$

The Mean Response: The mean response of Y given $\mathbf{x'} = \mathbf{x_0'} = (1, x_{10}, x_{20}, \ldots, x_{k0})$ is $\mu_{Y|x_{10}, x_{20}, \ldots, x_{k0}} = \beta_0 + \beta_1 x_{10} + \cdots + \beta_k x_{k0}$. The random variable $\hat{Y}_0 = \hat{\beta}_0 + \hat{\beta}_1 x_{10} + \cdots + \hat{\beta}_k x_{k0}$ is used to estimate

Table 4. Probability and Statistics Formulas (Continued)

$\mu_{Y|x_{10}, x_{20}, \ldots, x_{k0}}$:

1. $E(\hat{Y}_0) = \beta_0 + \beta_1 x_{10} + \cdots + \beta_k x_{k0}$

2. $\text{Var}(\hat{Y}_0) = \sigma^2 \mathbf{x}_0'(\mathbf{X}'\mathbf{X})^{-1}\mathbf{x}_0$

3. $\hat{Y}_0$ has a normal distribution.

4. $T = \dfrac{\hat{Y}_0 - \mu_{Y|x_{10}, x_{20}, \ldots, x_{k0}}}{S\sqrt{\mathbf{x}_0'(\mathbf{X}'\mathbf{X})^{-1}\mathbf{x}_0}}$ has a t distribution with $n - k - 1$ degrees of freedom.

5. A $100(1-\alpha)\%$ confidence interval for $\mu_{Y|x_{10}, x_{20}, \ldots, x_{k0}}$ has as endpoints $\hat{y}_0 \pm t_{\alpha/2, n-k-1} \cdot s\sqrt{\mathbf{x}_0'(\mathbf{X}'\mathbf{X})^{-1}\mathbf{x}_0}$.

6. Hypothesis test

Null Hypothesis	Alternative Hypothesis	Test Statistic	Rejection Region
$\beta_0 + \beta_1 x_{10} + \cdots + \beta_k x_{k0}$ $= y_0 = \mu_0$	$y_0 > \mu_0$ $y_0 < \mu_0$ $y_0 \neq \mu_0$	$T = \dfrac{\hat{Y}_0 - \mu_0}{S\sqrt{\mathbf{x}_0'(\mathbf{X}'\mathbf{X})^{-1}\mathbf{x}_0}}$	$T \geq t_{\alpha, n-k-1}$ $T \leq -t_{\alpha, n-k-1}$ $|T| \geq t_{\alpha/2, n-k-1}$

Prediction Interval: A prediction interval for a value y_0 of the random variable $Y_0 = \beta_0 + \beta_1 x_{10} + \cdots + \beta_k x_{k0} + \epsilon_0$ is obtained by considering the random variable $\hat{Y}_0 - Y_0$.

1. $E(\hat{Y}_0 - Y_0) = 0$

2. $\text{Var}(\hat{Y}_0 - Y_0) = \sigma^2 \left[1 + \mathbf{x}_0'(\mathbf{X}'\mathbf{X})^{-1}\mathbf{x}_0\right]$

3. $\hat{Y}_0 - Y_0$ has a normal distribution.

4. $T = \dfrac{\hat{Y}_0 - Y_0}{S\sqrt{1 + \mathbf{x}_0'(\mathbf{X}'\mathbf{X})^{-1}\mathbf{x}_0}}$ has a t distribution with $n - k - 1$ degrees of freedom.

5. A $100(1-\alpha)\%$ prediction interval for y_0 has as endpoints $\hat{y}_0 \pm t_{\alpha/2, n-k-1} \cdot s\sqrt{1 + \mathbf{x}_0'(\mathbf{X}'\mathbf{X})^{-1}\mathbf{x}_0}$

Analysis Of Variance Table:

Source of Variation	Sum of Squares	Degrees of Freedom	Mean Square	Computed F
Regression	SSR	k	$MSR = \frac{SSR}{k}$	MSR/MSE
Error	SSE	$n - k - 1$	$MSE = \frac{SSE}{n-k-1}$	
Total	SST	$n - 1$		

Hypothesis Test of Significant Regression:

Null Hypothesis	Alternative Hypothesis	Test Statistic	Rejection Region
$\beta_1 = \beta_2 = \cdots = \beta_k = 0$	$\beta_i \neq 0$ for some i	$F = MSR/MSE$	$F \geq F_{\alpha, k, n-k-1}$

Table 4. Probability and Statistics Formulas (Continued)

Sequential Sum Of Squares: Define

$$\mathbf{g} = \mathbf{X'y} = \begin{pmatrix} g_0 = \sum_{i=1}^{n} y_i \\ g_1 = \sum_{i=1}^{n} x_{1i} y_i \\ \vdots \\ g_k = \sum_{i=1}^{n} x_{ki} y_i \end{pmatrix}$$

$$SSR = \sum_{j=0}^{k} \hat{\beta}_j g_j - n\overline{y}^2$$

$SS(\beta_1, \beta_2, \ldots, \beta_r) = $ the sum of squares due to $\beta_1, \beta_2, \ldots, \beta_r$

$$= \sum_{j=1}^{r} \hat{\beta}_j g_j - n\overline{y}^2$$

$SS(\beta_1) = $ the regression sum of squares due to x_1

$$= \sum_{j=0}^{1} \hat{\beta}_j g_j - n\overline{y}^2$$

$SS(\beta_2 \mid \beta_1) = $ the regression sum of squares due to x_2 given x_1 is in the model

$$= SS(\beta_1, \beta_2) - SS(\beta_1) = \hat{\beta}_2 g_2$$

$SS(\beta_3 \mid \beta_1, \beta_2) = $ the regression sum of squares due to x_3 given x_1, x_2 are in the model

$$= SS(\beta_1, \beta_2, \beta_3) - SS(\beta_1, \beta_2) = \hat{\beta}_3 g_3$$

$$\vdots$$

$SS(\beta_r \mid \beta_1, \ldots, \beta_{r-1}) = $ the regression sum of squares due to x_r given $x_1, \ldots, x_{r-1}$ are in the model

$$= SS(\beta_1, \ldots, \beta_r) - SS(\beta_1, \ldots, \beta_{r-1}) = \hat{\beta}_r g_r$$

Partial F Test:

$Y_i = \beta_0 + \beta_1 x_{1i} + \cdots + \beta_r x_{ri} + \beta_{r+1} x_{(r+1)i} + \cdots + \beta_k x_{ki} + \epsilon_i$:   Full Model

$SSE(F) = $ sum of squares due to error in the full model

$Y_i = \beta_0 + \beta_1 x_{1i} + \cdots + \beta_r x_{ri} + \epsilon_i$:   Reduced Model

$SSE(R) = $ sum of squares due to error in the reduced model

$SS(\beta_{r+1}, \ldots, \beta_k \mid \beta_1, \ldots, \beta_r) = $ the regression sum of squares due to $x_{r+1}, \ldots, x_k$ given $x_1, \ldots, x_r$ are in the model

$$= SS(\beta_1, \ldots, \beta_r, \beta_{r+1}, \ldots, \beta_k) - SS(\beta_1, \ldots, \beta_r)$$

$$= \sum_{j=r+1}^{k} \hat{\beta}_j g_j$$

Null Hypothesis: $\beta_{r+1} = \beta_{r+2} = \cdots = \beta_k = 0$

Alternative Hypothesis: $\beta_m \neq 0$ for some $m = r+1, r+2, \ldots, k$

Test Statistic: $F = \dfrac{(SSE(R) - SSE(F))/(k-r)}{SSE(F)/(n-k-1)} = \dfrac{SS(\beta_{r+1}, \ldots, \beta_k \mid \beta_1, \ldots, \beta_r)/(k-r)}{SSE(F)/(n-k-1)}$

Rejection Region: $F \geq F_{\alpha, k-r, n-k-1}$

Table 4. Probability and Statistics Formulas (Continued)

Residual Analysis: Let h_{ii} be the diagonal entries of the HAT matrix given by $\mathbf{H} = \mathbf{X}(\mathbf{X}'\mathbf{X})^{-1}\mathbf{X}'$.

Standardized Residuals: $\dfrac{e_i}{\sqrt{MSE}} = \dfrac{e_i}{s}, \quad i = 1, 2, \ldots, n$

Studentized Residual: $e_i^* = \dfrac{e_i}{s\sqrt{1 - h_{ii}}}, \quad i = 1, 2, \ldots, n$

Deleted Studentized Residual: $d_i^* = e_i \left[\dfrac{n - k - 2}{s^2(1 - h_{ii}) - e_i^2} \right]^{1/2}, \quad i = 1, 2, \ldots, n$

Cook's Distance: $D_i = \dfrac{e_i^2}{(k+1)s^2} \left[\dfrac{h_{ii}}{(1 - h_{ii})^2} \right], \quad i = 1, 2, \ldots, n$

Press Residuals: $\delta_i = y_i - \hat{y}_{i,-i} = \dfrac{e_i}{1 - h_{ii}}, \quad i = 1, 2, \ldots, n$

where $\hat{y}_{i,-i}$ is the ith predicted value by the model without using the ith observation in calculating the regression coefficients.

Prediction Sum of Squares $= PRESS = \displaystyle\sum_{i=1}^{n} \delta_i^2$

$\displaystyle\sum_{i=1}^{n} |\, \delta_i \,|$ may also be used for cross validation. It is less sensitive to large press residuals.

The Analysis Of Variance

One-Way Anova

The Model: Let there be k independent random samples of size n_i, $i = 1, 2, \ldots, k$, $N = n_1 + n_2 + \cdots + n_k$, such that each population is normally distributed with mean μ_i and common variance σ^2. Let y_{ij} be the jth observation in the ith group, or treatment. Then

$$y_{ij} = \mu_i + e_{ij}$$

where e_{ij} is an observed value of the random error, ϵ_i. (Alternative model assumptions: the ϵ_i's are independent, normally distributed, with mean 0 and variance σ^2.)

Let α_i be the ith treatment effect and let μ be the grand mean. Then

$$y_{ij} = \mu + \alpha_i + e_{ij} \qquad \mu = \frac{\displaystyle\sum_{i=1}^{k} \mu_i}{k} \qquad \sum_{i=1}^{k} \alpha_i = 0$$

The Sum-Of-Squares Identity:

$\overline{y}_{i.}$ is the mean of the observations in the ith sample, $\overline{y}_{..}$ is the mean of all the observations.

$T_{i.}$ is the sum of all observations in the ith sample, $T_{..}$ is the sum of all N observations.

$$\underbrace{\sum_{i=1}^{k} \sum_{j=1}^{n_i} (y_{ij} - \overline{y}_{..})^2}_{SST} = \underbrace{\sum_{i=1}^{k} n_i (\overline{y}_{i.} - \overline{y}_{..})^2}_{SSA} + \underbrace{\sum_{i=1}^{k} \sum_{j=1}^{n_i} (y_{ij} - \overline{y}_{i.})^2}_{SSE}$$

$$SST = \text{the total sum of squares} = \sum_{i=1}^{k} \sum_{j=1}^{n_i} (y_{ij} - \overline{y}_{..})^2 = \sum_{i=1}^{k} \sum_{j=1}^{n_i} y_{ij}^2 - \frac{T_{..}^2}{N}$$

Table 4. Probability and Statistics Formulas (Continued)

$$SSA = \text{the sum of squares due to treatment} = \sum_{i=1}^{k} n_i (\overline{y}_{i.} - \overline{y}_{..})^2 = \sum_{i=1}^{k} \frac{T_{i.}^2}{n_i} - \frac{T_{...}^2}{N}$$

$$SSE = \text{the sum of squares due to error} = \sum_{i=1}^{k} \sum_{j=1}^{n_i} (y_{ij} - \overline{y}_{i.})^2 = SST - SSA$$

Properties:

1. $E[MSA] = E[S_A^2] = E\left[\dfrac{SSA}{k-1}\right] = \sigma^2 + \dfrac{\sum_{i=1}^{k} n_i \alpha_i^2}{k-1}$

2. $E[MSE] = E[S^2] = E\left[\dfrac{SSE}{N-k}\right] = \sigma^2$

3. $F = S_A^2/S^2$ has a F distribution with $k-1$ and $N-k$ degrees of freedom.

Analysis Of Variance Table:

Source of Variation	Sum of Squares	Degrees of Freedom	Mean Square	Computed F
Treatments	SSA	$k-1$	$s_A^2 = \frac{SSA}{k-1}$	s_A^2/s^2
Error	SSE	$N-k$	$s^2 = \frac{SSE}{N-k}$	
Total	SST	$N-1$		

Hypothesis Test:

Null Hypothesis: $\mu_1 = \mu_2 = \cdots = \mu_k$ $(\alpha_1 = \alpha_2 = \cdots = \alpha_k = 0)$

Alternative Hypothesis: at least two of the means are not equal $(\alpha_i \neq 0$ for some $i)$

Test Statistic: $F = S_A^2/S^2$

Rejection Region: $F \geq F_{\alpha, k-1, N-k}$

Multiple Comparison Procedures:

Tukey's Procedure:

Equal Sample Sizes:

Let $n = n_i$, $i = 1, 2, \ldots, k$ and let Q_{α, ν_1, ν_2} be a critical value of the Studentized Range distribution.

The set of confidence intervals with endpoints

$$(\overline{y}_{i.} - \overline{y}_{j.}) \pm Q_{\alpha, k, k(n-1)} \cdot \sqrt{s/n} \quad \text{for all } i \text{ and } j, i \neq j$$

is a collection of simultaneous $100(1-\alpha)\%$ confidence intervals for the differences between the true treatment means, $\mu_i - \mu_j$. Each confidence interval that does not include zero suggests $\mu_i \neq \mu_j$ at level α.

Unequal Sample Sizes:

The set of confidence intervals with endpoints

$$(\overline{y}_{i.} - \overline{y}_{j.}) \pm \frac{1}{\sqrt{2}} Q_{\alpha, k, N-k} \cdot s \sqrt{\frac{1}{n_i} + \frac{1}{n_j}} \quad \text{for all } i \text{ and } j, i \neq j$$

is a collection of simultaneous $100(1-\alpha)\%$ confidence intervals for the differences between the true treatment means, $\mu_i - \mu_j$.

Table 4. Probability and Statistics Formulas (Continued)

Duncan's Multiple Range Test:

Let $n = n_i$, $i = 1, 2, \ldots, k$ and let r_{α, ν_1, ν_2} be a critical value for Duncan's multiple range test. Duncan's procedure for determining significant differences between each treatment group at the joint significance level α is:

$$\text{Define} \quad R_p = r_{\alpha, p, k(n-1)} \cdot \sqrt{\frac{s^2}{n}} \quad \text{for } p = 2, 3, \ldots, k$$

List the sample means in increasing order. Compare the range of every subset of p, $p = 2, 3, \ldots, k$, sample means in the ordered list with R_p. If the range of a p-subset is less than R_p then that subset of ordered means is not significantly different.

Dunnett's Procedure:

Let $n = n_i$, $i = 0, 1, 2, \ldots, k$, d_{α, ν_1, ν_2} be a critical value for Dunnett's procedure, and let treatment 0 be the control group. Dunnett's procedure for determining significant differences between each treatment and the control at the joint significance level α is given by

Null Hypotheses: $\mu_0 = \mu_i$ $i = 1, 2, \ldots, k$

Alternative Hypotheses: $\mu_0 > \mu_i$
$$\mu_0 < \mu_i \qquad i = 1, 2, \ldots, k$$
$$\mu_0 \neq \mu_i$$

Test Statistics: $D_i = \dfrac{\overline{Y}_{i.} - \overline{Y}_{0.}}{\sqrt{2S^2/n}}$ $i = 1, 2, \ldots, k$

Rejection Region: $D_i \geq d_{\alpha, k, k(n-1)}$
$$D_i \leq -d_{\alpha, k, k(n-1)} \qquad i = 1, 2, \ldots, k$$
$$| D_i | \geq d_{\alpha/2, k, k(n-1)}$$

Contrast: A contrast L is a linear combination of the means μ_i such that the coefficients c_i sum to zero:

$$L = \sum_{i=1}^{k} c_i \mu_i \quad \text{where} \quad \sum_{i=1}^{k} c_i = 0$$

Let $\hat{L} = \sum_{i=1}^{k} c_i \overline{Y}_{i.}$, then

1. $\hat{L}$ has a normal distribution, $E(\hat{L}) = \sum_{i=1}^{k} c_i \mu_i$, $\text{Var}(\hat{L}) = \sigma^2 \sum_{i=1}^{k} \dfrac{c_i^2}{n_i}$

2. A $100(1 - \alpha)\%$ confidence interval for L has as endpoints

$$\hat{l} \pm t_{\alpha/2, N-k} \cdot s \sqrt{\sum_{i=1}^{k} c_i^2 / n_i}$$

3. Single degree of freedom test:

Null Hypothesis: $\sum_{i=1}^{k} c_i \mu_i = c$

Alternative Hypothesis: $\sum_{i=1}^{k} c_i \mu_i > c$, $\sum_{i=1}^{k} c_i \mu_i < c$, $\sum_{i=1}^{k} c_i \mu_i \neq c$

39

Table 4. Probability and Statistics Formulas (Continued)

Test Statistic: $T = \dfrac{L - c}{s\sqrt{\sum\limits_{i=1}^{k} c_i^2/n_i}}$ $\left(F = T^2 = \dfrac{(L - c)^2}{s^2 \sum\limits_{i=1}^{k} c_i^2/n_i} \right)$

Rejection Region: $T \geq t_{\alpha, N-k}, \quad T \leq -t_{\alpha, N-k}, \quad |T| \geq t_{\alpha/2, N-k}, \quad (F \geq F_{\alpha, 1, N-k})$

4. The set of confidence intervals with endpoints

$$\hat{l} \pm \sqrt{(k-1)F_{\alpha, k-1, N-k}} \cdot s \sqrt{\sum_{i=1}^{k} c_i^2/n_i}$$

is the collection of simultaneous $100(1 - \alpha)\%$ confidence intervals for all possible contrasts.

5. Let $n_i = n$, $i = 1, 2, \ldots, k$, then the contrast sum of squares, SSL, is given by

$$SSL = \frac{\left(\sum\limits_{i=1}^{k} c_i T_{i.} \right)^2}{n \sum\limits_{i=1}^{k} c_i^2}$$

6. Two contrasts $L_1 = \sum\limits_{i=1}^{k} b_i \mu_i$ and $L_2 = \sum\limits_{i=1}^{k} c_i \mu_i$ are orthogonal if $\sum\limits_{i=1}^{k} b_i c_i/n_i = 0$

Two-Way Anova (Completely Randomized Design or Randomized Complete Block Design)

The Model: Let there be a levels of factor A, b levels of factor B, and n treatment replications (ab cells, n observations in each cell, abn total observations). The observations in the (ij)th cell are assumed to be a random sample of size n from a normal population with mean μ_{ij} and variance σ^2. Let y_{ijk} be the kth observation at the ith level of factor A and the jth level of factor B. Then

$$y_{ijk} = \mu_{ij} + e_{ijk}$$

where e_{ijk} is an observed value of the random error, ϵ_{ijk}. (Alternative model assumptions: the ϵ_{ijk}'s are independent, normally distributed, with mean 0 and variance σ^2.)

Let α_i be the effect of the ith level of factor A, β_j be the effect of the jth level of factor B, $(\alpha\beta)_{ij}$ be the interaction effect of the ith level of factor A and the jth level of factor B, and μ be the grand mean. Then

$$y_{ijk} = \mu + \alpha_i + \beta_j + (\alpha\beta)_{ij} + e_{ijk}$$

where

$$\mu = \frac{\sum\limits_{i=1}^{a} \sum\limits_{j=1}^{b} \mu_{ij}}{ab}, \quad \sum_{i=1}^{a} \alpha_i = 0, \quad \sum_{j=1}^{b} \beta_j = 0, \quad \sum_{i=1}^{a} (\alpha\beta)_{ij} = \sum_{j=1}^{b} (\alpha\beta)_{ij} = 0$$

The Sum-Of-Squares Identity:

Dots in the subscript of $\bar{y}$ and T indicate the average and sum of y_{ijk}, respectively, over the appropriate subscript(s).

$$SST = SSA + SSB + SS(AB) + SSE$$

$$SST = \text{the total sum of squares} = \sum_{i=1}^{a} \sum_{j=1}^{b} \sum_{k=1}^{n} (y_{ijk} - \bar{y}_{...})^2 = \sum_{i=1}^{a} \sum_{j=1}^{b} \sum_{k=1}^{n} y_{ijk}^2 - \frac{T_{...}^2}{abn}$$

Table 4. Probability and Statistics Formulas (Continued)

$$SSA = \text{the sum of squares due to factor } A = bn\sum_{i=1}^{a}(\bar{y}_{i..} - \bar{y}_{...})^2 = \frac{\sum_{i=1}^{a} T_{i..}^2}{bn} - \frac{T_{...}^2}{abn}$$

$$SSB = \text{the sum of squares due to factor } B = an\sum_{j=1}^{b}(\bar{y}_{.j.} - \bar{y}_{...})^2 = \frac{\sum_{j=1}^{b} T_{.j.}^2}{an} - \frac{T_{...}^2}{abn}$$

$SS(AB) = $ the sum of squares due to interaction

$$= n\sum_{i=1}^{a}\sum_{j=1}^{b}(\bar{y}_{ij.} - \bar{y}_{i..} - \bar{y}_{.j.} + \bar{y}_{...})^2 = \frac{\sum_{i=1}^{a}\sum_{j=1}^{b} T_{ij.}^2}{n} - \frac{\sum_{i=1}^{a} T_{i..}^2}{bn} - \frac{\sum_{j=1}^{b} T_{.j.}^2}{an} + \frac{T_{...}^2}{abn}$$

$SSE = $ the sum of squares due to error

$$= \sum_{i=1}^{a}\sum_{j=1}^{b}\sum_{k=1}^{n}(y_{ijk} - \bar{y}_{ij.})^2 = SST - SSA - SSB - SS(AB)$$

Properties:

1. $E[MSA] = E[S_A^2] = E\left[\dfrac{SSA}{a-1}\right] = \sigma^2 + \dfrac{nb\sum_{i=1}^{a}\alpha_i^2}{a-1}$

2. $E[MSB] = E[S_B^2] = E\left[\dfrac{SSB}{b-1}\right] = \sigma^2 + \dfrac{na\sum_{j=1}^{b}\beta_j^2}{b-1}$

3. $E[MS(AB)] = E[S_{AB}^2] = E\left[\dfrac{SS(AB)}{(a-1)(b-1)}\right] = \sigma^2 + \dfrac{n\sum_{i=1}^{a}\sum_{j=1}^{b}(\alpha\beta)_{ij}^2}{(a-1)(b-1)}$

4. $E[MSE] = E[S^2] = E\left[\dfrac{SSE}{ab(n-1)}\right] = \sigma^2$

5. $F = S_A^2/S^2$ has an F distribution with $a-1$ and $ab(n-1)$ degrees of freedom.

   $F = S_B^2/S^2$ has an F distribution with $b-1$ and $ab(n-1)$ degrees of freedom.

   $F = S_{AB}^2/S^2$ has an F distribution with $(a-1)(b-1)$ and $ab(n-1)$ degrees of freedom.

Analysis Of Variance Table:

Source of Variation	Sum of Squares	Degrees of Freedom	Mean Square	Computed F
Factor A	SSA	$a-1$	$s_A^2 = \frac{SSA}{a-1}$	s_A^2/s^2
Factor B	SSB	$b-1$	$s_B^2 = \frac{SSB}{b-1}$	s_B^2/s^2
Interaction AB	$SS(AB)$	$(a-1)(b-1)$	$s_{AB}^2 = \frac{SS(AB)}{(a-1)(b-1)}$	s_{AB}^2/s^2
Error	SSE	$ab(n-1)$	$s^2 = \frac{SSE}{ab(n-1)}$	
Total	SST	$abn-1$		

Table 4. Probability and Statistics Formulas (Continued)

Hypothesis Tests:

Null Hypothesis	Alternative Hypothesis	Test Statistic	Rejection Region
$\alpha_1 = \cdots = \alpha_a = 0$	$\alpha_i \neq 0$ for some i	$F = S_A^2/S^2$	$F \geq F_{\alpha, a-1, ab(n-1)}$
$\beta_1 = \cdots = \beta_b = 0$	$\beta_j \neq 0$ for some j	$F = S_B^2/S^2$	$F \geq F_{\alpha, b-1, ab(n-1)}$
$(\alpha\beta)_{11} = \cdots = (\alpha\beta)_{ab} = 0$	$(\alpha\beta)_{ij} \neq 0$ for some (ij)	$F = S_{AB}^2/S^2$	$F \geq F_{\alpha, (a-1)(b-1), ab(n-1)}$

Three-Way Anova (Completely Randomized Design or Randomized Complete Block Design)

The Model: Let there be three factors A, B, and C, with levels a, b, and c, respectively, and n treatment replications. The observations in the (ijk)th cell are assumed to be from a random sample of size n from a normal population with mean μ_{ijk} and variance σ^2. Let y_{ijkl} be the lth observation at the ith level of factor A, the jth level of factor B, and the kth level of factor C, and let e_{ijkl} be an observed value of the random error, ϵ_{ijkl}. (Alternative model assumptions: the ϵ_{ijkl}'s are independent, normally distributed, with mean 0 and variance σ^2.) Then

$$y_{ijkl} = \mu_{ijk} + e_{ijkl}$$
$$= \mu + \alpha_i + \beta_j + \gamma_k + (\alpha\beta)_{ij} + (\alpha\gamma)_{ik} + (\beta\gamma)_{jk} + (\alpha\beta\gamma)_{ijk} + e_{ijkl}$$

$$\mu = \text{the grand mean} = \frac{\sum_{i=1}^{a}\sum_{j=1}^{b}\sum_{k=1}^{c}\mu_{ijk}}{abc}$$

α_i, β_j, γ_k = main effects

$$\sum_{i=1}^{a}\alpha_i = 0, \quad \sum_{j=1}^{b}\beta_j = 0, \quad \sum_{k=1}^{c}\gamma_k = 0$$

$(\alpha\beta)_{ij}$, $(\alpha\gamma)_{ik}$, $(\beta\gamma)_{jk}$ = two-factor interaction effects

$$\sum_{i=1}^{a}(\alpha\beta)_{ij} = \sum_{j=1}^{b}(\alpha\beta)_{ij} = \sum_{i=1}^{a}(\alpha\gamma)_{ik} = \sum_{k=1}^{c}(\alpha\gamma)_{ik} = \sum_{j=1}^{b}(\beta\gamma)_{jk} = \sum_{k=1}^{c}(\beta\gamma)_{jk} = 0$$

$(\alpha\beta\gamma)_{ijk}$ = three-factor interaction effect

$$\sum_{i=1}^{a}(\alpha\beta\gamma)_{ijk} = \sum_{j=1}^{b}(\alpha\beta\gamma)_{ijk} = \sum_{k=1}^{c}(\alpha\beta\gamma)_{ijk} = 0$$

The Sum-Of-Squares Identity:

Dots in the subscript of $\overline{y}$ and T indicate the average and sum of y_{ijkl}, respectively, over the appropriate subscript(s).

$$SST = SSA + SSB + SSC + SS(AB) + SS(AC) + SS(BC) + SS(ABC) + SSE$$

$$SST = \text{the total sum of squares} = \sum_{i=1}^{a}\sum_{j=1}^{b}\sum_{k=1}^{c}\sum_{l=1}^{n}(y_{ijkl} - \overline{y}_{....})^2 = \sum_{i=1}^{a}\sum_{j=1}^{b}\sum_{k=1}^{c}\sum_{l=1}^{n}y_{ijkl}^2 - \frac{T_{....}^2}{abcn}$$

$$SSA = \text{the sum of squares due to factor } A = bcn\sum_{i=1}^{a}(\overline{y}_{i...} - \overline{y}_{....})^2 = \frac{\sum_{i=1}^{a}T_{i...}^2}{bcn} - \frac{T_{....}^2}{abcn}$$

Table 4. Probability and Statistics Formulas (Continued)

$$SSB = \text{the sum of squares due to factor } B = acn\sum_{j=1}^{b}(\overline{y}_{.j..} - \overline{y}_{....})^2 = \frac{\sum\limits_{j=1}^{b} T^2_{.j..}}{acn} - \frac{T^2_{....}}{abcn}$$

$$SSC = \text{the sum of squares due to factor } C = abn\sum_{k=1}^{c}(\overline{y}_{..k.} - \overline{y}_{....})^2 = \frac{\sum\limits_{k=1}^{c} T^2_{..k.}}{abn} - \frac{T^2_{....}}{abcn}$$

$SS(AB) = \text{the sum of squares due to interaction between factor } A \text{ and factor } B$

$$= cn\sum_{i=1}^{a}\sum_{j=1}^{b}(\overline{y}_{ij..} - \overline{y}_{i...} - \overline{y}_{.j..} + \overline{y}_{....})^2 = \frac{\sum\limits_{i=1}^{a}\sum\limits_{j=1}^{b} T^2_{ij..}}{cn} - \frac{\sum\limits_{i=1}^{a} T^2_{i...}}{bcn} - \frac{\sum\limits_{j=1}^{b} T^2_{.j..}}{acn} + \frac{T^2_{....}}{abcn}$$

$SS(AC) = \text{the sum of squares due to interaction between factor } A \text{ and factor } C$

$$= bn\sum_{i=1}^{a}\sum_{k=1}^{c}(\overline{y}_{i.k.} - \overline{y}_{i...} - \overline{y}_{..k.} + \overline{y}_{....})^2 = \frac{\sum\limits_{i=1}^{a}\sum\limits_{k=1}^{c} T^2_{i.k.}}{bn} - \frac{\sum\limits_{i=1}^{a} T^2_{i...}}{bcn} - \frac{\sum\limits_{k=1}^{c} T^2_{..k.}}{abn} + \frac{T^2_{....}}{abcn}$$

$SS(BC) = \text{the sum of squares due to interaction between factor } B \text{ and factor } C$

$$= an\sum_{j=1}^{b}\sum_{k=1}^{c}(\overline{y}_{.jk.} - \overline{y}_{.j..} - \overline{y}_{..k.} + \overline{y}_{....})^2 = \frac{\sum\limits_{j=1}^{b}\sum\limits_{k=1}^{c} T^2_{.jk.}}{an} - \frac{\sum\limits_{j=1}^{b} T^2_{.j..}}{acn} - \frac{\sum\limits_{k=1}^{c} T^2_{..k.}}{abn} + \frac{T^2_{....}}{abcn}$$

$SS(ABC) = \text{the sum of squares due to interaction between factors } A, B, \text{ and } C$

$$= \sum_{i=1}^{a}\sum_{j=1}^{b}\sum_{k=1}^{c}(\overline{y}_{ijk.} - \overline{y}_{ij..} - \overline{y}_{i.k.} - \overline{y}_{.jk.} + \overline{y}_{i...} + \overline{y}_{.j..} + \overline{y}_{..k.} - \overline{y}_{....})^2$$

$$= \frac{\sum\limits_{i=1}^{a}\sum\limits_{j=1}^{b}\sum\limits_{k=1}^{c} T^2_{ijk.}}{n} - \frac{\sum\limits_{i=1}^{a}\sum\limits_{j=1}^{b} T^2_{ij..}}{cn} - \frac{\sum\limits_{i=1}^{a}\sum\limits_{k=1}^{c} T^2_{i.k.}}{bn} - \frac{\sum\limits_{j=1}^{b}\sum\limits_{k=1}^{c} T^2_{.jk.}}{an} + \frac{\sum\limits_{i=1}^{a} T^2_{i...}}{bcn} + \frac{\sum\limits_{j=1}^{b} T^2_{.j..}}{acn} + \frac{\sum\limits_{k=1}^{c} T^2_{..k.}}{abn} - \frac{T^2_{....}}{abcn}$$

$SSE = \text{the sum of squares due to error}$

$$= \sum_{i=1}^{a}\sum_{j=1}^{b}\sum_{k=1}^{c}\sum_{l=1}^{n}(y_{ijkl} - \overline{y}_{ijk.})^2$$

$$= SST - SSA - SSB - SSC - SS(AB) - SS(AC) - SS(BC) - SS(ABC)$$

Properties:

1. $E[MSA] = E[S_A^2] = E\left[\dfrac{SSA}{a-1}\right] = \sigma^2 + \dfrac{bcn\sum\limits_{i=1}^{a}\alpha_i^2}{a-1}$

2. $E[MSB] = E[S_B^2] = E\left[\dfrac{SSB}{b-1}\right] = \sigma^2 + \dfrac{acn\sum\limits_{j=1}^{b}\beta_j^2}{b-1}$

3. $E[MSC] = E[S_C^2] = E\left[\dfrac{SSC}{c-1}\right] = \sigma^2 + \dfrac{abn\sum\limits_{k=1}^{c}\gamma_k^2}{c-1}$

Table 4. Probability and Statistics Formulas (Continued)

4. $E[MS(AB)] = E[S_{AB}^2] = E\left[\dfrac{SS(AB)}{(a-1)(b-1)}\right] = \sigma^2 + \dfrac{cn \sum\limits_{i=1}^{a} \sum\limits_{j=1}^{b} (\alpha\beta)_{ij}^2}{(a-1)(b-1)}$

5. $E[MS(AC)] = E[S_{AC}^2] = E\left[\dfrac{SS(AC)}{(a-1)(c-1)}\right] = \sigma^2 + \dfrac{bn \sum\limits_{i=1}^{a} \sum\limits_{k=1}^{c} (\alpha\gamma)_{ik}^2}{(a-1)(c-1)}$

6. $E[MS(BC)] = E[S_{BC}^2] = E\left[\dfrac{SS(BC)}{(b-1)(c-1)}\right] = \sigma^2 + \dfrac{an \sum\limits_{j=1}^{b} \sum\limits_{k=1}^{c} (\beta\gamma)_{jk}^2}{(b-1)(c-1)}$

7. $E[MS(ABC)] = E[S_{ABC}^2] = E\left[\dfrac{SS(ABC)}{(a-1)(b-1)(c-1)}\right] = \sigma^2 + \dfrac{n \sum\limits_{i=1}^{a} \sum\limits_{j=1}^{b} \sum\limits_{k=1}^{c} (\alpha\beta\gamma)_{ijk}^2}{(a-1)(b-1)(c-1)}$

8. $E[MSE] = E[S^2] = E\left[\dfrac{SSE}{abc(n-1)}\right] = \sigma^2$

Analysis Of Variance Table:

Source of Variation	Sum of Squares	Degrees of Freedom	Mean Square	Computed F
Factor A	SSA	$a-1$	$s_A^2 = \frac{SSA}{a-1}$	s_A^2/s^2
Factor B	SSB	$b-1$	$s_B^2 = \frac{SSB}{b-1}$	s_B^2/s^2
Factor C	SSC	$c-1$	$s_C^2 = \frac{SSC}{c-1}$	s_C^2/s^2
Interaction AB	$SS(AB)$	$(a-1)(b-1)$	$s_{AB}^2 = \frac{SS(AB)}{(a-1)(b-1)}$	s_{AB}^2/s^2
Interaction AC	$SS(AC)$	$(a-1)(c-1)$	$s_{AC}^2 = \frac{SS(AC)}{(a-1)(c-1)}$	s_{AC}^2/s^2
Interaction BC	$SS(BC)$	$(b-1)(c-1)$	$s_{BC}^2 = \frac{SS(BC)}{(b-1)(c-1)}$	s_{BC}^2/s^2
Interaction ABC	$SS(ABC)$	$(a-1)(b-1)(c-1)$	$s_{ABC}^2 = \frac{SS(ABC)}{(a-1)(b-1)(c-1)}$	s_{ABC}^2/s^2
Error	SSE	$abc(n-1)$	$s^2 = \frac{SSE}{abc(n-1)}$	
Total	SST	$abcn-1$		

Hypothesis Tests:

Null Hypothesis	Alternative Hypothesis	Test Statistic	Rejection Region
$\alpha_1 = \cdots = \alpha_a = 0$	$\alpha_i \neq 0$, some i	$F = S_A^2/S^2$	$F \geq F_{\alpha,a-1,abc(n-1)}$
$\beta_1 = \cdots = \beta_b = 0$	$\beta_j \neq 0$, some j	$F = S_B^2/S^2$	$F \geq F_{\alpha,b-1,abc(n-1)}$
$\gamma_1 = \cdots = \gamma_c = 0$	$\gamma_k \neq 0$, some k	$F = S_C^2/S^2$	$F \geq F_{\alpha,c-1,abc(n-1)}$
$(\alpha\beta)_{11} = \cdots = (\alpha\beta)_{ab} = 0$	$(\alpha\beta)_{ij} \neq 0$, some (ij)	$F = S_{AB}^2/S^2$	$F \geq F_{\alpha,(a-1)(b-1),abc(n-1)}$
$(\alpha\gamma)_{11} = \cdots = (\alpha\gamma)_{ac} = 0$	$(\alpha\gamma)_{ik} \neq 0$, some (ik)	$F = S_{AC}^2/S^2$	$F \geq F_{\alpha,(a-1)(c-1),abc(n-1)}$
$(\beta\gamma)_{11} = \cdots = (\beta\gamma)_{bc} = 0$	$(\beta\gamma)_{jk} \neq 0$, some (jk)	$F = S_{BC}^2/S^2$	$F \geq F_{\alpha,(b-1)(c-1),abc(n-1)}$
$(\alpha\beta\gamma)_{111} = \cdots = (\alpha\beta\gamma)_{abc} = 0$	$(\alpha\beta\gamma)_{ijk} \neq 0$, some (ijk)	$F = S_{ABC}^2/S^2$	$F \geq F_{\alpha,(a-1)(b-1)(c-1),abc(n-1)}$

Table 4. Probability and Statistics Formulas (Continued)

Latin Squares

The Model: In an $r \times r$ Latin Square, let $y_{ij(k)}$ be an observation from a normal population with mean $\mu_{ij(k)}$ and variance σ^2 corresponding to the ith row, jth column, and kth treatment. (The parentheses in the subscripts are used to denote the one value k assumes for each (i, j) combination, $i, j, k = 1, 2, \ldots, r$). Then

$$y_{ij(k)} = \mu + \alpha_i + \beta_j + \tau_k + e_{ij(k)}$$

$$\mu = \frac{\displaystyle\sum_{i=1}^{r} \sum_{j=1}^{r} \mu_{ij(k)}}{r^2} = \text{the grand mean}$$

$e_{ij(k)}$ is an observed value of the random error $\epsilon_{ij(k)}$. (Alternative model assumptions: the $\epsilon_{ij(k)}$'s are independent, normally distributed with mean 0 and variance σ^2.)

$\alpha_i, \; \beta_j, \; \tau_k,$ are the row, column, and treatment effects, respectively, and

$$\sum_{i=1}^{r} \alpha_i = \sum_{j=1}^{r} \beta_j = \sum_{k=1}^{r} \tau_k = 0$$

Sum-Of-Squares Identity:

Dots in the subscript of $\bar{y}$ and T indicate the average and sum of $y_{ij(k)}$, respectively, over the appropriate subscript(s).

$$SST = SSR + SSC + SSTr + SSE$$

$$SST = \text{the total sum of squares} = \sum_{i=1}^{r} \sum_{j=1}^{r} (y_{ij(k)} - \bar{y}_{...})^2 = \sum_{i=1}^{r} \sum_{j=1}^{r} y_{ij(k)}^2 - \frac{T_{...}^2}{r^2}$$

$$SSR = \text{the sum of squares due to rows} = r \sum_{i=1}^{r} (\bar{y}_{i..} - \bar{y}_{...})^2 = \frac{\displaystyle\sum_{i=1}^{r} T_{i..}^2}{r} - \frac{T_{...}^2}{r^2}$$

$$SSC = \text{the sum of squares due to columns} = r \sum_{j=1}^{r} (\bar{y}_{.j.} - \bar{y}_{...})^2 = \frac{\displaystyle\sum_{j=1}^{r} T_{.j.}^2}{r} - \frac{T_{...}^2}{r^2}$$

$$SSTr = \text{the sum of squares due to treatment} = r \sum_{k=1}^{r} (\bar{y}_{..k} - \bar{y}_{...})^2 = \frac{\displaystyle\sum_{k=1}^{r} T_{..k}^2}{r} - \frac{T_{...}^2}{r^2}$$

$$SSE = \text{the sum of squares due to error} = \sum_{i=1}^{r} \sum_{j=1}^{r} (y_{ij(k)} - \bar{y}_{i..} - \bar{y}_{.j.} - \bar{y}_{..k} + 2\bar{y}_{...})^2$$

$$= SST - SSR - SSC - SSTr$$

Properties:

1. $E[MSR] = E[S_R^2] = E\left[\dfrac{SSR}{r-1}\right] = \sigma^2 + \dfrac{r \displaystyle\sum_{i=1}^{r} \alpha_i^2}{r-1}$

2. $E[MSC] = E[S_C^2] = E\left[\dfrac{SSC}{r-1}\right] = \sigma^2 + \dfrac{r \displaystyle\sum_{j=1}^{r} \beta_j^2}{r-1}$

45

Table 4. Probability and Statistics Formulas (Continued)

$$E[S_{Tr}^2] = E\left[\frac{SSTr}{r-1}\right] = \sigma^2 + \frac{r\sum_{k=1}^{r}\tau_k^2}{r-1}$$

4. $E[SSE] = E[S^2] = E\left[\dfrac{SSE}{(r-1)(r-2)}\right] = \sigma^2$

Analysis Of Variance Table:

Source of Variation	Sum of Squares	Degrees of Freedom	Mean Square	Computed F
Rows	SSR	$r-1$	$s_R^2 = \frac{SSR}{r-1}$	s_R^2/s^2
Columns	SSC	$r-1$	$s_C^2 = \frac{SSC}{r-1}$	s_C^2/s^2
Treatments	$SSTr$	$r-1$	$s_{Tr}^2 = \frac{SSTr}{r-1}$	s_{Tr}^2/s^2
Error	SSE	$(r-1)(r-2)$	$s^2 = \frac{SSE}{(r-1)(r-2)}$	
Total	SST	$r^2 - 1$		

Hypothesis Tests:

Null Hypothesis	Alternative Hypothesis	Test Statistic	Rejection Region
$\alpha_1 = \cdots = \alpha_r = 0$	$\alpha_i \neq 0$ for some i	$F = S_R^2/S^2$	$F \geq F_{\alpha,r-1,(r-1)(r-2)}$
$\beta_1 = \cdots = \beta_r = 0$	$\beta_j \neq 0$ for some j	$F = S_C^2/S^2$	$F \geq F_{\alpha,r-1,(r-1)(r-2)}$
$\tau_1 = \cdots = \tau_r = 0$	$\tau_k \neq 0$ for some k	$F = S_{Tr}^2/S^2$	$F \geq F_{\alpha,r-1,(r-1)(r-2)}$

Nonparametric Statistics

The Sign Test

Assumptions: Let $X_1, X_2, \ldots, X_n$ be a random sample from a continuous distribution.

Hypothesis Test:

Null Hypothesis: $\tilde{\mu} = \tilde{\mu}_0$

Alternative Hypothesis: $\tilde{\mu} > \tilde{\mu}_0$, $\quad \tilde{\mu} < \tilde{\mu}_0$, $\quad \tilde{\mu} \neq \tilde{\mu}_0$

Test Statistic: $Y =$ the number of X_i's greater than $\tilde{\mu}_0$.
Under the null hypothesis, Y has a binomial distribution with parameters n and $p = .5$.

Rejection Region: $Y \geq c_1$, $\quad Y \leq c_2$, $\quad Y \geq c$ or $Y \leq n - c$
The critical values c_1, c_2, and c are obtained from the binomial distribution with parameters n and $p = .5$ to yield the desired significance level α.

Sample values equal to $\tilde{\mu}_0$ are excluded from the analysis and the sample size is reduced accordingly.

The Normal Approximation: When $n \geq 10$ and $p = .5$ the binomial distribution can be approximated by a normal distribution with

$\mu_Y = np = .5n \quad$ and $\quad \sigma_Y^2 = np(1-p) = .25n$

$Z = \dfrac{Y - .5n}{.5\sqrt{n}}$ has approximately a standard normal distribution when H_0 is true and $n \geq 10$.

The Wilcoxon Signed-Rank Test

Assumptions: Let $X_1, X_2, \ldots, X_n$ be a random sample from a continuous distribution.

Table 4. Probability and Statistics Formulas (Continued)

Hypothesis Test:

Null Hypothesis: $\tilde{\mu} = \tilde{\mu}_0$

Alternative Hypothesis: $\tilde{\mu} > \tilde{\mu}_0, \quad \tilde{\mu} < \tilde{\mu}_0, \quad \tilde{\mu} \neq \tilde{\mu}_0$

Rank the absolute differences $\mid X_1 - \tilde{\mu}_0 \mid, \mid X_2 - \tilde{\mu}_0 \mid, \ldots, \mid X_n - \tilde{\mu}_0 \mid$.

Test Statistic: $T_+ =$ the sum of the ranks corresponding to the positive differences $(X_i - \tilde{\mu}_0)$.

Rejection Region: $T_+ \geq c_1, \quad T_+ \leq c_2, \quad T_+ \geq c$ or $T_+ \leq n(n+1) - c$

$c_1, c_2,$ and c are critical values for the Wilcoxon Signed-Rank Statistic such that $P(T_+ \geq c_1) \approx \alpha$, $P(T_+ \leq c_2) \approx \alpha$, and $P(T_+ \geq c) \approx \alpha/2$.

Any observed difference $(x_i - \tilde{\mu}_0) = 0$ is excluded from the test and the sample size is reduced accordingly.

The Normal Approximation: When $n \geq 20$, T_+ has approximately a normal distribution with

$$\mu_{T_+} = \frac{n(n+1)}{4} \quad \text{and} \quad \sigma^2_{T_+} = \frac{n(n+1)(2n+1)}{24}$$

$Z = \dfrac{T_+ - \mu_{T_+}}{\sigma_{T_+}}$ has approximately a standard normal distribution when H_0 is true.

The Wilcoxon Rank-Sum (Mann-Whitney) Test

Assumptions: Let $X_1, X_2, \ldots, X_m$ and $Y_1, Y_2, \ldots, Y_n$, $m \leq n$, be independent random samples from continuous distributions.

Hypothesis Test:

Null Hypothesis: $\tilde{\mu}_1 - \tilde{\mu}_2 = \Delta_0$

Alternative Hypothesis: $\tilde{\mu}_1 - \tilde{\mu}_2 > \Delta_0, \quad \tilde{\mu}_1 - \tilde{\mu}_2 < \Delta_0, \quad \tilde{\mu}_1 - \tilde{\mu}_2 \neq \Delta_0$

Subtract Δ_0 from each X_i. Combine the $(X_i - \Delta_0)$'s and the Y_j's into one sample and rank all of the observations.

Test Statistic: $W = \displaystyle\sum_{i=1}^{m} R_i$, where R_i is the rank of $(X_i - \Delta_0)$ in the combined sample.

Rejection Region: $W \geq c_1, \quad W \leq c_2, \quad W \geq c$ or $W \leq m(m+n+1) - c$

$c_1, c_2,$ and c are critical values for the Wilcoxon rank-sum statistic such that $P(W \geq c_1) \approx \alpha$, $P(W \leq c_2) \approx \alpha$, and $P(W \geq c) \approx \alpha/2$.

The Normal Approximation: When both m and n are greater than 8, W has approximately a normal distribution with

$$\mu_W = \frac{m(m+n+1)}{2} \quad \text{and} \quad \sigma^2_W = \frac{mn(m+n+1)}{12}$$

$Z = \dfrac{W - \mu_W}{\sigma_W}$ has approximately a standard normal distribution.

The Mann-Whitney U Statistic: The rank-sum test can also be based on the statistic

$$U = W - \frac{m(n+1)}{2}$$

When both m and n are greater than 8, U has approximately a normal distribution with

$$\mu_U = \frac{mn}{2} \quad \text{and} \quad \sigma^2_U = \frac{mn(m+n+1)}{12}$$

Table 4. Probability and Statistics Formulas (Continued)

$Z = \dfrac{U - \mu_U}{\sigma_U}$ has approximately a standard normal distribution.

The Kruskal-Wallis Test

Assumptions: Let there be $k > 2$ independent random samples from continuous distributions, n_i, $i = 1, 2, \ldots, k$, be the number of observations in each sample, and $n = n_1 + n_2 + \cdots n_k$.

Hypothesis Test:

Null Hypothesis: the k samples are from identical populations.

Alternative Hypothesis: at least two of the populations differ in location.

Rank all n observations from 1 (smallest) to n (largest). Let R_i be the total of the ranks in the ith sample.

Test Statistic: $H = \dfrac{12}{n(n+1)} \displaystyle\sum_{i=1}^{k} \dfrac{R_i^2}{n_i} - 3(n+1)$

If H_0 is true and either
1. $k = 3$, $n_i \geq 6$, $i = 1, 2, 3$ or
2. $k > 3$, $n_i \geq 5$, $i = 1, 2, \ldots, k$
then H has a chi-square distribution with $k - 1$ degrees of freedom,

Rejection Region: $H \geq \chi^2_{\alpha, k-1}$

The Friedman F_r Test For A Randomized Block Design

Assumptions: Let there be k independent random samples (treatments) from continuous distributions and b blocks.

Hypothesis Test:

Null Hypothesis: the k samples are from identical populations.

Alternative Hypothesis: at least two of the populations differ in location.

Rank each observation from 1 (smallest) to k (largest) within each block. Let R_i be the rank sum of the ith sample (treatment).

Test Statistic: $F_r = \dfrac{12}{bk(k+1)} \displaystyle\sum_{i=1}^{k} R_i^2 - 3b(k+1)$

Rejection Region: $F_r \geq \chi^2_{\alpha, k-1}$

The Runs Test

Run: a run is a maximal subsequence of elements with a common property.

Hypothesis Test:

Null Hypothesis: the sequence is random

Alternative Hypothesis: the sequence is not random

Test Statistic: $V =$ the total number of runs

Rejection Region: $V \geq v_1$ or $V \leq v_2$
v_1 and v_2 are critical values for the runs test such that $P(V \geq v_1) \approx \alpha/2$ and $P(V \leq v_2) \approx \alpha/2$.

The Normal Approximation: Let m be the number of elements with the property that occurs least and n be the number of elements with the other property. As m and n increase, V has approximately a normal distribution with

Table 4. Probability and Statistics Formulas (Continued)

$$\mu_V = \frac{2mn}{m+n} + 1 \quad \text{and} \quad \sigma_V^2 = \frac{2mn(2mn - m - n)}{(m+n)^2(m+n-1)}$$

$Z = \dfrac{V - \mu_V}{\sigma_V}$ has approximately a standard normal distribution when H_0 is true.

Spearman's Rank Correlation Coefficient:

Let there be n pairs of observations from the continuous distributions X and Y. Rank the observations in the two samples separately from smallest to largest. Let u_i be the rank of the ith observation in the first sample and let v_i be the rank of the ith observation in the second sample. Spearman's rank correlation coefficient, r_S, is a measure of the correlation between ranks, calculated by using the ranks in place of the actual observations in the formula for the correlation coefficient r.

$$r_S = \frac{SS_{uv}}{\sqrt{SS_{uu}SS_{vv}}} = \frac{n\sum\limits_{i=1}^{n} u_i v_i - \left(\sum\limits_{i=1}^{n} u_i\right)\left(\sum\limits_{i=1}^{n} v_i\right)}{\sqrt{\left[n\sum\limits_{i=1}^{n} u_i^2 - \left(\sum\limits_{i=1}^{n} u_i\right)^2\right]\left[n\sum\limits_{i=1}^{n} v_i^2 - \left(\sum\limits_{i=1}^{n} v_i\right)^2\right]}}$$

$$= 1 - \frac{6\sum\limits_{i=1}^{n} d_i^2}{n(n^2 - 1)} \quad \text{where} \quad d_i = u_i - v_i.$$

The shortcut formula for r_S that uses the differences d_i, $i = 1, 2, \ldots, n$, is not exact when there are tied measurements. The approximation is good when the number of ties is small in comparison to n.

Hypothesis Test:

Null Hypothesis: $\rho_S = 0$ (no population correlation between ranks)

Alternative Hypothesis: $\rho_S > 0, \quad \rho_S < 0, \quad \rho_S \neq 0$

Test Statistic: r_S

Rejection Region: $r_S \geq r_{S,\alpha}, \quad r_S \leq -r_{S,\alpha}, \quad |r_S| \geq r_{S,\alpha/2}$

The Normal Approximation: When H_0 is true r_S has approximately a normal distribution with

$$\mu_{r_S} = 0 \quad \text{and} \quad \sigma_{r_S}^2 = \frac{1}{n-1}$$

$Z = \dfrac{r_S - 0}{1/\sqrt{n-1}} = r_S\sqrt{n-1}$ has approximately a standard normal distribution as n increases.

Table 5. The Binomial Cumulative Distribution Function

Let X be a binomial random variable characterized by the parameters n and p. This table contains values of the binomial cumulative distribution function $B(x; n, p) = P(X \le x) = \sum_{y=0}^{x} b(y; n, p) = \sum_{y=0}^{x} \binom{n}{y} p^y (1-p)^{n-y}$.

$n = 5$

x	.01	.05	.10	.20	.25	.30	.40	.50	.60	.70	.75	.80	.90	.95	.99
0	.9510	.7738	.5905	.3277	.2373	.1681	.0778	.0313	.0102	.0024	.0010	.0003	.0000		
1	.9990	.9774	.9185	.7373	.6328	.5282	.3370	.1875	.0870	.0308	.0156	.0067	.0005	.0000	
2	1.0000	.9988	.9914	.9421	.8965	.8369	.6826	.5000	.3174	.1631	.1035	.0579	.0086	.0012	.0000
3		1.0000	.9995	.9933	.9844	.9692	.9130	.8125	.6630	.4718	.3672	.2627	.0815	.0226	.0010
4			1.0000	.9997	.9990	.9976	.9898	.9688	.9222	.8319	.7627	.6723	.4095	.2262	.0490

$n = 10$

x	.01	.05	.10	.20	.25	.30	.40	.50	.60	.70	.75	.80	.90	.95	.99
0	.9044	.5987	.3487	.1074	.0563	.0282	.0060	.0010	.0001	.0000					
1	.9957	.9139	.7361	.3758	.2440	.1493	.0464	.0107	.0017	.0001	.0000	.0000			
2	.9999	.9885	.9298	.6778	.5256	.3828	.1673	.0547	.0123	.0016	.0004	.0001	.0000		
3	1.0000	.9990	.9872	.8791	.7759	.6496	.3823	.1719	.0548	.0106	.0035	.0009	.0000		
4		.9999	.9984	.9672	.9219	.8497	.6331	.3770	.1662	.0473	.0197	.0064	.0001	.0000	
5		1.0000	.9999	.9936	.9803	.9527	.8338	.6230	.3669	.1503	.0781	.0328	.0016	.0001	
6			1.0000	.9991	.9965	.9894	.9452	.8281	.6177	.3504	.2241	.1209	.0128	.0010	.0000
7				.9999	.9996	.9984	.9877	.9453	.8327	.6172	.4744	.3222	.0702	.0115	.0001
8				1.0000	1.0000	.9999	.9983	.9893	.9536	.8507	.7560	.6242	.2639	.0861	.0043
9						1.0000	.9999	.9990	.9940	.9718	.9437	.8926	.6513	.4013	.0956

$n = 15$

x	.01	.05	.10	.20	.25	.30	.40	.50	.60	.70	.75	.80	.90	.95	.99
0	.8601	.4633	.2059	.0352	.0134	.0047	.0005	.0000							
1	.9904	.8290	.5490	.1671	.0802	.0353	.0052	.0005	.0000						
2	.9996	.9638	.8159	.3980	.2361	.1268	.0271	.0037	.0003	.0000					
3	1.0000	.9945	.9444	.6482	.4613	.2969	.0905	.0176	.0019	.0001	.0000				
4		.9994	.9873	.8358	.6865	.5155	.2173	.0592	.0093	.0007	.0001	.0000			
5		.9999	.9978	.9389	.8516	.7216	.4032	.1509	.0338	.0037	.0008	.0001			
6		1.0000	.9997	.9819	.9434	.8689	.6098	.3036	.0950	.0152	.0042	.0008			
7			1.0000	.9958	.9827	.9500	.7869	.5000	.2131	.0500	.0173	.0042	.0000		
8				.9992	.9958	.9848	.9050	.6964	.3902	.1311	.0566	.0181	.0003	.0000	
9				.9999	.9992	.9963	.9662	.8491	.5968	.2784	.1484	.0611	.0022	.0001	
10				1.0000	.9999	.9993	.9907	.9408	.7827	.4845	.3135	.1642	.0127	.0006	
11					1.0000	.9999	.9981	.9824	.9095	.7031	.5387	.3518	.0556	.0055	.0000
12						1.0000	.9997	.9963	.9729	.8732	.7639	.6020	.1841	.0362	.0004
13							1.0000	.9995	.9948	.9647	.9198	.8329	.4510	.1710	.0096
14								1.0000	.9995	.9953	.9866	.9648	.7941	.5367	.1399

Table 5. The Binomial Cumulative Distribution Function (Continued)

$n = 20$

x	.01	.05	.10	.20	.25	.30	.40	.50	.60	.70	.75	.80	.90	.95	.99
0	.8179	.3585	.1216	.0115	.0032	.0008	.0000								
1	.9831	.7358	.3917	.0692	.0243	.0076	.0005	.0000							
2	.9990	.9245	.6769	.2061	.0913	.0355	.0036	.0002							
3	1.0000	.9841	.8670	.4114	.2252	.1071	.0160	.0013	.0000						
4		.9974	.9568	.6296	.4148	.2375	.0510	.0059	.0003						
5		.9997	.9887	.8042	.6172	.4164	.1256	.0207	.0016	.0000					
6		1.0000	.9976	.9133	.7858	.6080	.2500	.0577	.0065	.0003	.0000				
7			.9996	.9679	.8982	.7723	.4159	.1316	.0210	.0013	.0002	.0000			
8			.9999	.9900	.9591	.8867	.5956	.2517	.0565	.0051	.0009	.0001			
9			1.0000	.9974	.9861	.9520	.7553	.4119	.1275	.0171	.0039	.0006			
10				.9994	.9961	.9829	.8725	.5881	.2447	.0480	.0139	.0026	.0000		
11				.9999	.9991	.9949	.9435	.7483	.4044	.1133	.0409	.0100	.0001		
12				1.0000	.9998	.9987	.9790	.8684	.5841	.2277	.1018	.0321	.0004		
13					1.0000	.9997	.9935	.9423	.7500	.3920	.2142	.0867	.0024	.0000	
14						1.0000	.9984	.9793	.8744	.5836	.3828	.1958	.0113	.0003	
15							.9997	.9941	.9490	.7625	.5852	.3704	.0432	.0026	
16							1.0000	.9987	.9840	.8929	.7748	.5886	.1330	.0159	.0000
17								.9998	.9964	.9645	.9087	.7939	.3231	.0755	.0010
18								1.0000	.9995	.9924	.9757	.9308	.6083	.2642	.0169
19									1.0000	.9992	.9968	.9885	.8784	.6415	.1821

$n = 25$

x	.01	.05	.10	.20	.25	.30	.40	.50	.60	.70	.75	.80	.90	.95	.99
0	.7778	.2774	.0718	.0038	.0008	.0001	.0000								
1	.9742	.6424	.2712	.0274	.0070	.0016	.0001								
2	.9980	.8729	.5371	.0982	.0321	.0090	.0004	.0000							
3	.9999	.9659	.7636	.2340	.0962	.0332	.0024	.0001							
4	1.0000	.9928	.9020	.4207	.2137	.0905	.0095	.0005	.0000						
5		.9988	.9666	.6167	.3783	.1935	.0294	.0020	.0001						
6		.9998	.9905	.7800	.5611	.3407	.0736	.0073	.0003						
7		1.0000	.9977	.8909	.7265	.5118	.1536	.0216	.0012	.0000					
8			.9995	.9532	.8506	.6769	.2735	.0539	.0043	.0001					
9			.9999	.9827	.9287	.8106	.4246	.1148	.0132	.0005	.0000				
10			1.0000	.9944	.9703	.9022	.5858	.2122	.0344	.0018	.0002	.0000			
11				.9985	.9893	.9558	.7323	.3450	.0778	.0060	.0009	.0001			
12				.9996	.9966	.9825	.8462	.5000	.1538	.0175	.0034	.0004			
13				.9999	.9991	.9940	.9222	.6550	.2677	.0442	.0107	.0015			
14				1.0000	.9998	.9982	.9656	.7878	.4142	.0978	.0297	.0056	.0000		
15					1.0000	.9995	.9868	.8852	.5754	.1894	.0713	.0173	.0001		
16						.9999	.9957	.9461	.7265	.3231	.1494	.0468	.0005		
17						1.0000	.9988	.9784	.8464	.4882	.2735	.1091	.0023	.0000	
18							.9997	.9927	.9264	.6593	.4389	.2200	.0095	.0002	
19							.9999	.9980	.9706	.8065	.6217	.3833	.0334	.0012	
20							1.0000	.9995	.9905	.9095	.7863	.5793	.0980	.0072	.0000
21								.9999	.9976	.9668	.9038	.7660	.2364	.0341	.0001
22								1.0000	.9996	.9910	.9679	.9018	.4629	.1271	.0020
23									.9999	.9984	.9930	.9726	.7288	.3576	.0258
24									1.0000	.9999	.9992	.9962	.9282	.7226	.2222

Table 6. The Poisson Cumulative Distribution Function

Let X be a Poisson random variable characterized by the parameter μ. This table contains values of the Poisson cumulative distribution function $F(x;\mu) = P(X \le x) = \sum_{y=0}^{x} \frac{e^{-\mu}\mu^y}{y!}$.

x	.05	.10	.15	.20	.25	μ .30	.35	.40	.45	.50
0	.9512	.9048	.8607	.8187	.7788	.7408	.7047	.6703	.6376	.6065
1	.9988	.9953	.9898	.9825	.9735	.9631	.9513	.9384	.9246	.9098
2	1.0000	.9998	.9995	.9989	.9978	.9964	.9945	.9921	.9891	.9856
3		1.0000	1.0000	.9999	.9999	.9997	.9995	.9992	.9988	.9982
4				1.0000	1.0000	1.0000	1.0000	.9999	.9999	.9998
5								1.0000	1.0000	1.0000

x	.55	.60	.65	.70	.75	μ .80	.85	.90	.95	1.00
0	.5769	.5488	.5220	.4966	.4724	.4493	.4274	.4066	.3867	.3679
1	.8943	.8781	.8614	.8442	.8266	.8088	.7907	.7725	.7541	.7358
2	.9815	.9769	.9717	.9659	.9595	.9526	.9451	.9371	.9287	.9197
3	.9975	.9966	.9956	.9942	.9927	.9909	.9889	.9865	.9839	.9810
4	.9997	.9996	.9994	.9992	.9989	.9986	.9982	.9977	.9971	.9963
5	1.0000	1.0000	.9999	.9999	.9999	.9998	.9997	.9997	.9995	.9994
6		1.0000	1.0000	1.0000	1.0000	1.0000	1.0000	1.0000	.9999	.9999
7									1.0000	1.0000

x	1.1	1.2	1.3	1.4	1.5	μ 1.6	1.7	1.8	1.9	2.0
0	.3329	.3012	.2725	.2466	.2231	.2019	.1827	.1653	.1496	.1353
1	.6990	.6626	.6268	.5918	.5578	.5249	.4932	.4628	.4337	.4060
2	.9004	.8795	.8571	.8335	.8088	.7834	.7572	.7306	.7037	.6767
3	.9743	.9662	.9569	.9463	.9344	.9212	.9068	.8913	.8747	.8571
4	.9946	.9923	.9893	.9857	.9814	.9763	.9704	.9636	.9559	.9473
5	.9990	.9985	.9978	.9968	.9955	.9940	.9920	.9896	.9868	.9834
6	.9999	.9997	.9996	.9994	.9991	.9987	.9981	.9974	.9966	.9955
7	1.0000	1.0000	.9999	.9999	.9998	.9997	.9996	.9994	.9992	.9989
8			1.0000	1.0000	1.0000	1.0000	.9999	.9999	.9998	.9998
9						1.0000	1.0000	1.0000	1.0000	1.0000

Table 6. The Poisson Cumulative Distribution Function (Continued)

x	2.1	2.2	2.3	2.4	2.5	μ 2.6	2.7	2.8	2.9	3.0
0	.1225	.1108	.1003	.0907	.0821	.0743	.0672	.0608	.0550	.0498
1	.3796	.3546	.3309	.3084	.2873	.2674	.2487	.2311	.2146	.1991
2	.6496	.6227	.5960	.5697	.5438	.5184	.4936	.4695	.4460	.4232
3	.8386	.8194	.7993	.7787	.7576	.7360	.7141	.6919	.6696	.6472
4	.9379	.9275	.9162	.9041	.8912	.8774	.8629	.8477	.8318	.8153
5	.9796	.9751	.9700	.9643	.9580	.9510	.9433	.9349	.9258	.9161
6	.9941	.9925	.9906	.9884	.9858	.9828	.9794	.9756	.9713	.9665
7	.9985	.9980	.9974	.9967	.9958	.9947	.9934	.9919	.9901	.9881
8	.9997	.9995	.9994	.9991	.9989	.9985	.9981	.9976	.9969	.9962
9	.9999	.9999	.9999	.9998	.9997	.9996	.9995	.9993	.9991	.9989
10	1.0000	1.0000	1.0000	1.0000	.9999	.9999	.9999	.9998	.9998	.9997
11				1.0000	1.0000	1.0000	1.0000	1.0000	.9999	.9999
12									1.0000	1.0000

x	3.1	3.2	3.3	3.4	3.5	μ 3.6	3.7	3.8	3.9	4.0
0	.0450	.0408	.0369	.0334	.0302	.0273	.0247	.0224	.0202	.0183
1	.1847	.1712	.1586	.1468	.1359	.1257	.1162	.1074	.0992	.0916
2	.4012	.3799	.3594	.3397	.3208	.3027	.2854	.2689	.2531	.2381
3	.6248	.6025	.5803	.5584	.5366	.5152	.4942	.4735	.4532	.4335
4	.7982	.7806	.7626	.7442	.7254	.7064	.6872	.6678	.6484	.6288
5	.9057	.8946	.8829	.8705	.8576	.8441	.8301	.8156	.8006	.7851
6	.9612	.9554	.9490	.9421	.9347	.9267	.9182	.9091	.8995	.8893
7	.9858	.9832	.9802	.9769	.9733	.9692	.9648	.9599	.9546	.9489
8	.9953	.9943	.9931	.9917	.9901	.9883	.9863	.9840	.9815	.9786
9	.9986	.9982	.9978	.9973	.9967	.9960	.9952	.9942	.9931	.9919
10	.9996	.9995	.9994	.9992	.9990	.9987	.9984	.9981	.9977	.9972
11	.9999	.9999	.9998	.9998	.9997	.9996	.9995	.9994	.9993	.9991
12	1.0000	1.0000	1.0000	.9999	.9999	.9999	.9999	.9998	.9998	.9997
13				1.0000	1.0000	1.0000	1.0000	1.0000	.9999	.9999
14									1.0000	1.0000

Table 6. The Poisson Cumulative Distribution Function (Continued)

x	5	6	7	8	9	μ 10	15	20	25	30
0	.0067	.0025	.0009	.0003	.0001	.0000				
1	.0404	.0174	.0073	.0030	.0012	.0005				
2	.1247	.0620	.0296	.0138	.0062	.0028	.0000			
3	.2650	.1512	.0818	.0424	.0212	.0103	.0002			
4	.4405	.2851	.1730	.0996	.0550	.0293	.0009	.0000		
5	.6160	.4457	.3007	.1912	.1157	.0671	.0028	.0001		
6	.7622	.6063	.4497	.3134	.2068	.1301	.0076	.0003		
7	.8666	.7440	.5987	.4530	.3239	.2202	.0180	.0008	.0000	
8	.9319	.8472	.7291	.5925	.4557	.3328	.0374	.0021	.0001	
9	.9682	.9161	.8305	.7166	.5874	.4579	.0699	.0050	.0002	
10	.9863	.9574	.9015	.8159	.7060	.5830	.1185	.0108	.0006	.0000
11	.9945	.9799	.9467	.8881	.8030	.6968	.1848	.0214	.0014	.0001
12	.9980	.9912	.9730	.9362	.8758	.7916	.2676	.0390	.0031	.0002
13	.9993	.9964	.9872	.9658	.9261	.8645	.3632	.0661	.0065	.0004
14	.9998	.9986	.9943	.9827	.9585	.9165	.4657	.1049	.0124	.0009
15	.9999	.9995	.9976	.9918	.9780	.9513	.5681	.1565	.0223	.0019
16	1.0000	.9998	.9990	.9963	.9889	.9730	.6641	.2211	.0377	.0039
17		.9999	.9996	.9984	.9947	.9857	.7489	.2970	.0605	.0073
18		1.0000	.9999	.9993	.9976	.9928	.8195	.3814	.0920	.0129
19			1.0000	.9997	.9989	.9965	.8752	.4703	.1336	.0219
20				.9999	.9996	.9984	.9170	.5591	.1855	.0353
21				1.0000	.9998	.9993	.9469	.6437	.2473	.0544
22					.9999	.9997	.9673	.7206	.3175	.0806
23					1.0000	.9999	.9805	.7875	.3939	.1146
24						1.0000	.9888	.8432	.4734	.1572
25							.9938	.8878	.5529	.2084
26							.9967	.9221	.6294	.2673
27							.9983	.9475	.7002	.3329
28							.9991	.9657	.7634	.4031
29							.9996	.9782	.8179	.4757
30							.9998	.9865	.8633	.5484
31							.9999	.9919	.8999	.6186
32							1.0000	.9953	.9285	.6845
33								.9973	.9502	.7444
34								.9985	.9662	.7973
35								.9992	.9775	.8426
36								.9996	.9854	.8804
37								.9998	.9908	.9110
38								.9999	.9943	.9352
39								.9999	.9966	.9537
40								1.0000	.9980	.9677
41									.9988	.9779
42									.9993	.9852
43									.9996	.9903
44									.9998	.9937

Table 7. Cumulative Distribution Function for the Standard Normal Random Variable

This table contains values of the cumulative distribution function for the standard normal random variable $\Phi(z) =$
$P(Z \le z) = \int\limits_{-\infty}^{z} \frac{1}{\sqrt{2\pi}} e^{-z^2/2} \, dz.$

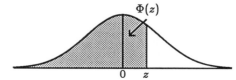

z	.00	.01	.02	.03	.04	.05	.06	.07	.08	.09
-3.4	.0003	.0003	.0003	.0003	.0003	.0003	.0003	.0003	.0003	.0002
-3.3	.0005	.0005	.0005	.0004	.0004	.0004	.0004	.0004	.0004	.0003
-3.2	.0007	.0007	.0006	.0006	.0006	.0006	.0006	.0005	.0005	.0005
-3.1	.0010	.0009	.0009	.0009	.0008	.0008	.0008	.0008	.0007	.0007
-3.0	.0013	.0013	.0013	.0012	.0012	.0011	.0011	.0011	.0010	.0010
-2.9	.0019	.0018	.0018	.0017	.0016	.0016	.0015	.0015	.0014	.0014
-2.8	.0026	.0025	.0024	.0023	.0023	.0022	.0021	.0021	.0020	.0019
-2.7	.0035	.0034	.0033	.0032	.0031	.0030	.0029	.0028	.0027	.0026
-2.6	.0047	.0045	.0044	.0043	.0041	.0040	.0039	.0038	.0037	.0036
-2.5	.0062	.0060	.0059	.0057	.0055	.0054	.0052	.0051	.0049	.0048
-2.4	.0082	.0080	.0078	.0075	.0073	.0071	.0069	.0068	.0066	.0064
-2.3	.0107	.0104	.0102	.0099	.0096	.0094	.0091	.0089	.0087	.0084
-2.2	.0139	.0136	.0132	.0129	.0125	.0122	.0119	.0116	.0113	.0110
-2.1	.0179	.0174	.0170	.0166	.0162	.0158	.0154	.0150	.0146	.0143
-2.0	.0228	.0222	.0217	.0212	.0207	.0202	.0197	.0192	.0188	.0183
-1.9	.0287	.0281	.0274	.0268	.0262	.0256	.0250	.0244	.0239	.0233
-1.8	.0359	.0351	.0344	.0336	.0329	.0322	.0314	.0307	.0301	.0294
-1.7	.0446	.0436	.0427	.0418	.0409	.0401	.0392	.0384	.0375	.0367
-1.6	.0548	.0537	.0526	.0516	.0505	.0495	.0485	.0475	.0465	.0455
-1.5	.0668	.0655	.0643	.0630	.0618	.0606	.0594	.0582	.0571	.0559
-1.4	.0808	.0793	.0778	.0764	.0749	.0735	.0721	.0708	.0694	.0681
-1.3	.0968	.0951	.0934	.0918	.0901	.0885	.0869	.0853	.0838	.0823
-1.2	.1151	.1131	.1112	.1093	.1075	.1056	.1038	.1020	.1003	.0985
-1.1	.1357	.1335	.1314	.1292	.1271	.1251	.1230	.1210	.1190	.1170
-1.0	.1587	.1562	.1539	.1515	.1492	.1469	.1446	.1423	.1401	.1379
-0.9	.1841	.1814	.1788	.1762	.1736	.1711	.1685	.1660	.1635	.1611
-0.8	.2119	.2090	.2061	.2033	.2005	.1977	.1949	.1922	.1894	.1867
-0.7	.2420	.2389	.2358	.2327	.2296	.2266	.2236	.2206	.2177	.2148
-0.6	.2743	.2709	.2676	.2643	.2611	.2578	.2546	.2514	.2483	.2451
-0.5	.3085	.3050	.3015	.2981	.2946	.2912	.2877	.2843	.2810	.2776
-0.4	.3446	.3409	.3372	.3336	.3300	.3264	.3228	.3192	.3156	.3121
-0.3	.3821	.3783	.3745	.3707	.3669	.3632	.3594	.3557	.3520	.3483
-0.2	.4207	.4168	.4129	.4090	.4052	.4013	.3974	.3936	.3897	.3859
-0.1	.4602	.4562	.4522	.4483	.4443	.4404	.4364	.4325	.4286	.4247
-0.0	.5000	.4960	.4920	.4880	.4840	.4801	.4761	.4721	.4681	.4641

Table 7. Cumulative Distribution Function for the Standard Normal Random Variable (Continued)

z	.00	.01	.02	.03	.04	.05	.06	.07	.08	.09
0.0	.5000	.5040	.5080	.5120	.5160	.5199	.5239	.5279	.5319	.5359
0.1	.5398	.5438	.5478	.5517	.5557	.5596	.5636	.5675	.5714	.5753
0.2	.5793	.5832	.5871	.5910	.5948	.5987	.6026	.6064	.6103	.6141
0.3	.6179	.6217	.6255	.6293	.6331	.6368	.6406	.6443	.6480	.6517
0.4	.6554	.6591	.6628	.6664	.6700	.6736	.6772	.6808	.6844	.6879
0.5	.6915	.6950	.6985	.7019	.7054	.7088	.7123	.7157	.7190	.7224
0.6	.7257	.7291	.7324	.7357	.7389	.7422	.7454	.7486	.7517	.7549
0.7	.7580	.7611	.7642	.7673	.7704	.7734	.7764	.7794	.7823	.7852
0.8	.7881	.7910	.7939	.7967	.7995	.8023	.8051	.8078	.8106	.8133
0.9	.8159	.8186	.8212	.8238	.8264	.8289	.8315	.8340	.8365	.8389
1.0	.8413	.8438	.8461	.8485	.8508	.8531	.8554	.8577	.8599	.8621
1.1	.8643	.8665	.8686	.8708	.8729	.8749	.8770	.8790	.8810	.8830
1.2	.8849	.8869	.8888	.8907	.8925	.8944	.8962	.8980	.8997	.9015
1.3	.9032	.9049	.9066	.9082	.9099	.9115	.9131	.9147	.9162	.9177
1.4	.9192	.9207	.9222	.9236	.9251	.9265	.9279	.9292	.9306	.9319
1.5	.9332	.9345	.9357	.9370	.9382	.9394	.9406	.9418	.9429	.9441
1.6	.9452	.9463	.9474	.9484	.9495	.9505	.9515	.9525	.9535	.9545
1.7	.9554	.9564	.9573	.9582	.9591	.9599	.9608	.9616	.9625	.9633
1.8	.9641	.9649	.9656	.9664	.9671	.9678	.9686	.9693	.9699	.9706
1.9	.9713	.9719	.9726	.9732	.9738	.9744	.9750	.9756	.9761	.9767
2.0	.9772	.9778	.9783	.9788	.9793	.9798	.9803	.9808	.9812	.9817
2.1	.9821	.9826	.9830	.9834	.9838	.9842	.9846	.9850	.9854	.9857
2.2	.9861	.9864	.9868	.9871	.9875	.9878	.9881	.9884	.9887	.9890
2.3	.9893	.9896	.9898	.9901	.9904	.9906	.9909	.9911	.9913	.9916
2.4	.9918	.9920	.9922	.9925	.9927	.9929	.9931	.9932	.9934	.9936
2.5	.9938	.9940	.9941	.9943	.9945	.9946	.9948	.9949	.9951	.9952
2.6	.9953	.9955	.9956	.9957	.9959	.9960	.9961	.9962	.9963	.9964
2.7	.9965	.9966	.9967	.9968	.9969	.9970	.9971	.9972	.9973	.9974
2.8	.9974	.9975	.9976	.9977	.9977	.9978	.9979	.9979	.9980	.9981
2.9	.9981	.9982	.9982	.9983	.9984	.9984	.9985	.9985	.9986	.9986
3.0	.9987	.9987	.9987	.9988	.9988	.9989	.9989	.9989	.9990	.9990
3.1	.9990	.9991	.9991	.9991	.9992	.9992	.9992	.9992	.9993	.9993
3.2	.9993	.9993	.9994	.9994	.9994	.9994	.9994	.9995	.9995	.9995
3.3	.9995	.9995	.9995	.9996	.9996	.9996	.9996	.9996	.9996	.9997
3.4	.9997	.9997	.9997	.9997	.9997	.9997	.9997	.9997	.9997	.9998

Critical Values, $P(Z \geq z_\alpha) = \alpha$

α	.10	.05	.025	.01	.005	.001	.0005	.0001	
z_α	1.2816	1.6449	1.9600	2.3263	2.5758	3.0902	3.2905	3.7190	

α	.00009	.00008	.00007	.00006	.00005	.00004	.00003	.00002	.00001
z_α	3.7455	3.7750	3.8082	3.8461	3.8906	3.9444	4.0128	4.1075	4.2649

Table 8. Critical Values For The t Distribution

This table contains critical values $t_{\alpha,\nu}$ for the t distribution defined by $P(T \geq t_{\alpha,\nu}) = \alpha$.

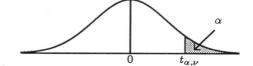

ν	.20	.10	.05	.025	.01	α .005	.001	.0005	.0001
1	1.3764	3.0777	6.3138	12.7062	31.8205	63.6567	318.3088	636.6192	3183.0988
2	1.0607	1.8856	2.9200	4.3027	6.9646	9.9248	22.3271	31.5991	70.7001
3	.9785	1.6377	2.3534	3.1824	4.5407	5.8409	10.2145	12.9240	22.2037
4	.9410	1.5332	2.1318	2.7764	3.7469	4.6041	7.1732	8.6103	13.0337
5	.9195	1.4759	2.0150	2.5706	3.3649	4.0321	5.8934	6.8688	9.6776
6	.9057	1.4398	1.9432	2.4469	3.1427	3.7074	5.2076	5.9588	8.0248
7	.8960	1.4149	1.8946	2.3646	2.9980	3.4995	4.7853	5.4079	7.0634
8	.8889	1.3968	1.8595	2.3060	2.8965	3.3554	4.5008	5.0413	6.4420
9	.8834	1.3830	1.8331	2.2622	2.8214	3.2498	4.2968	4.7809	6.0101
10	.8791	1.3722	1.8125	2.2281	2.7638	3.1693	4.1437	4.5869	5.6938
11	.8755	1.3634	1.7959	2.2010	2.7181	3.1058	4.0247	4.4370	5.4528
12	.8726	1.3562	1.7823	2.1788	2.6810	3.0545	3.9296	4.3178	5.2633
13	.8702	1.3502	1.7709	2.1604	2.6503	3.0123	3.8520	4.2208	5.1106
14	.8681	1.3450	1.7613	2.1448	2.6245	2.9768	3.7874	4.1405	4.9850
15	.8662	1.3406	1.7531	2.1314	2.6025	2.9467	3.7328	4.0728	4.8800
16	.8647	1.3368	1.7459	2.1199	2.5835	2.9208	3.6862	4.0150	4.7909
17	.8633	1.3334	1.7396	2.1098	2.5669	2.8982	3.6458	3.9651	4.7144
18	.8620	1.3304	1.7341	2.1009	2.5524	2.8784	3.6105	3.9216	4.6480
19	.8610	1.3277	1.7291	2.0930	2.5395	2.8609	3.5794	3.8834	4.5899
20	.8600	1.3253	1.7247	2.0860	2.5280	2.8453	3.5518	3.8495	4.5385
21	.8591	1.3232	1.7207	2.0796	2.5176	2.8314	3.5271	3.8192	4.4929
22	.8583	1.3212	1.7171	2.0739	2.5083	2.8187	3.5050	3.7921	4.4520
23	.8575	1.3195	1.7139	2.0687	2.4999	2.8073	3.4850	3.7676	4.4152
24	.8569	1.3178	1.7109	2.0639	2.4922	2.7969	3.4668	3.7454	4.3819
25	.8562	1.3163	1.7081	2.0595	2.4851	2.7874	3.4502	3.7251	4.3517
26	.8557	1.3150	1.7056	2.0555	2.4786	2.7787	3.4350	3.7066	4.3240
27	.8551	1.3137	1.7033	2.0518	2.4727	2.7707	3.4210	3.6896	4.2987
28	.8546	1.3125	1.7011	2.0484	2.4671	2.7633	3.4081	3.6739	4.2754
29	.8542	1.3114	1.6991	2.0452	2.4620	2.7564	3.3962	3.6594	4.2539
30	.8538	1.3104	1.6973	2.0423	2.4573	2.7500	3.3852	3.6460	4.2340
40	.8507	1.3031	1.6839	2.0211	2.4233	2.7045	3.3069	3.5510	4.0942
50	.8489	1.2987	1.6759	2.0086	2.4033	2.6778	3.2614	3.4960	4.0140
60	.8477	1.2958	1.6706	2.0003	2.3901	2.6603	3.2317	3.4602	3.9621
120	.8446	1.2886	1.6577	1.9799	2.3578	2.6174	3.1595	3.3735	3.8372
∞	.8416	1.2816	1.6449	1.9600	2.3263	2.5758	3.0902	3.2905	3.7190

Table 9. Critical Values For The Chi-Square Distribution

This table contains critical values $\chi^2_{\alpha,\nu}$ for the Chi-Square distribution defined by $P(\chi^2 \geq \chi^2_{\alpha,\nu}) = \alpha$.

ν	.9999	.9995	.999	.995	.99	.975	.95	.90
1	$.0^7157$	$.0^6393$	$.0^5157$	$.0^4393$	.0002	.0010	.0039	.0158
2	.0002	.0010	.0020	.0100	.0201	.0506	.1026	.2107
3	.0052	.0153	.0243	.0717	.1148	.2158	.3518	.5844
4	.0284	.0639	.0908	.2070	.2971	.4844	.7107	1.0636
5	.0822	.1581	.2102	.4117	.5543	.8312	1.1455	1.6103
6	.1724	.2994	.3811	.6757	.8721	1.2373	1.6354	2.2041
7	.3000	.4849	.5985	.9893	1.2390	1.6899	2.1673	2.8331
8	.4636	.7104	.8571	1.3444	1.6465	2.1797	2.7326	3.4895
9	.6608	.9717	1.1519	1.7349	2.0879	2.7004	3.3251	4.1682
10	.8889	1.2650	1.4787	2.1559	2.5582	3.2470	3.9403	4.8652
11	1.1453	1.5868	1.8339	2.6032	3.0535	3.8157	4.5748	5.5778
12	1.4275	1.9344	2.2142	3.0738	3.5706	4.4038	5.2260	6.3038
13	1.7333	2.3051	2.6172	3.5650	4.1069	5.0088	5.8919	7.0415
14	2.0608	2.6967	3.0407	4.0747	4.6604	5.6287	6.5706	7.7895
15	2.4082	3.1075	3.4827	4.6009	5.2293	6.2621	7.2609	8.5468
16	2.7739	3.5358	3.9416	5.1422	5.8122	6.9077	7.9616	9.3122
17	3.1567	3.9802	4.4161	5.6972	6.4078	7.5642	8.6718	10.0852
18	3.5552	4.4394	4.9048	6.2648	7.0149	8.2307	9.3905	10.8649
19	3.9683	4.9123	5.4068	6.8440	7.6327	8.9065	10.1170	11.6509
20	4.3952	5.3981	5.9210	7.4338	8.2604	9.5908	10.8508	12.4426
21	4.8348	5.8957	6.4467	8.0337	8.8972	10.2829	11.5913	13.2396
22	5.2865	6.4045	6.9830	8.6427	9.5425	10.9823	12.3380	14.0415
23	5.7494	6.9237	7.5292	9.2604	10.1957	11.6886	13.0905	14.8480
24	6.2230	7.4527	8.0849	9.8862	10.8564	12.4012	13.8484	15.6587
25	6.7066	7.9910	8.6493	10.5197	11.5240	13.1197	14.6114	16.4734
26	7.1998	8.5379	9.2221	11.1602	12.1981	13.8439	15.3792	17.2919
27	7.7019	9.0932	9.8028	11.8076	12.8785	14.5734	16.1514	18.1139
28	8.2126	9.6563	10.3909	12.4613	13.5647	15.3079	16.9279	18.9392
29	8.7315	10.2268	10.9861	13.1211	14.2565	16.0471	17.7084	19.7677
30	9.2581	10.8044	11.5880	13.7867	14.9535	16.7908	18.4927	20.5992
31	9.7921	11.3887	12.1963	14.4578	15.6555	17.5387	19.2806	21.4336
32	10.3331	11.9794	12.8107	15.1340	16.3622	18.2908	20.0719	22.2706
33	10.8810	12.5763	13.4309	15.8153	17.0735	19.0467	20.8665	23.1102
34	11.4352	13.1791	14.0567	16.5013	17.7891	19.8063	21.6643	23.9523
35	11.9957	13.7875	14.6878	17.1918	18.5089	20.5694	22.4650	24.7967
36	12.5622	14.4012	15.3241	17.8867	19.2327	21.3359	23.2686	25.6433
37	13.1343	15.0202	15.9653	18.5858	19.9602	22.1056	24.0749	26.4921
38	13.7120	15.6441	16.6112	19.2889	20.6914	22.8785	24.8839	27.3430
39	14.2950	16.2729	17.2616	19.9959	21.4262	23.6543	25.6954	28.1958
40	14.8831	16.9062	17.9164	20.7065	22.1643	24.4330	26.5093	29.0505
50	21.0093	23.4610	24.6739	27.9907	29.7067	32.3574	34.7643	37.6886
60	27.4969	30.3405	31.7383	35.5345	37.4849	40.4817	43.1880	46.4589
70	34.2607	37.4674	39.0364	43.2752	45.4417	48.7576	51.7393	55.3289
80	41.2445	44.7910	46.5199	51.1719	53.5401	57.1532	60.3915	64.2778
90	48.4087	52.2758	54.1552	59.1963	61.7541	65.6466	69.1260	73.2911
100	55.7246	59.8957	61.9179	67.3276	70.0649	74.2219	77.9295	82.3581

Table 9. Critical Values For The Chi-Square Distribution (Continued)

ν	.10	.05	.025	.01	.005	.001	.0005	.0001
1	2.7055	3.8415	5.0239	6.6349	7.8794	10.8276	12.1157	15.1367
2	4.6052	5.9915	7.3778	9.2103	10.5966	13.8155	15.2018	18.4207
3	6.2514	7.8147	9.3484	11.3449	12.8382	16.2662	17.7300	21.1075
4	7.7794	9.4877	11.1433	13.2767	14.8603	18.4668	19.9974	23.5127
5	9.2364	11.0705	12.8325	15.0863	16.7496	20.5150	22.1053	25.7448
6	10.6446	12.5916	14.4494	16.8119	18.5476	22.4577	24.1028	27.8563
7	12.0170	14.0671	16.0128	18.4753	20.2777	24.3219	26.0178	29.8775
8	13.3616	15.5073	17.5345	20.0902	21.9550	26.1245	27.8680	31.8276
9	14.6837	16.9190	19.0228	21.6660	23.5894	27.8772	29.6658	33.7199
10	15.9872	18.3070	20.4832	23.2093	25.1882	29.5883	31.4198	35.5640
11	17.2750	19.6751	21.9200	24.7250	26.7568	31.2641	33.1366	37.3670
12	18.5493	21.0261	23.3367	26.2170	28.2995	32.9095	34.8213	39.1344
13	19.8119	22.3620	24.7356	27.6882	29.8195	34.5282	36.4778	40.8707
14	21.0641	23.6848	26.1189	29.1412	31.3193	36.1233	38.1094	42.5793
15	22.3071	24.9958	27.4884	30.5779	32.8013	37.6973	39.7188	44.2632
16	23.5418	26.2962	28.8454	31.9999	34.2672	39.2524	41.3081	45.9249
17	24.7690	27.5871	30.1910	33.4087	35.7185	40.7902	42.8792	47.5664
18	25.9894	28.8693	31.5264	34.8053	37.1565	42.3124	44.4338	49.1894
19	27.2036	30.1435	32.8523	36.1909	38.5823	43.8202	45.9731	50.7955
20	28.4120	31.4104	34.1696	37.5662	39.9968	45.3147	47.4985	52.3860
21	29.6151	32.6706	35.4789	38.9322	41.4011	46.7970	49.0108	53.9620
22	30.8133	33.9244	36.7807	40.2894	42.7957	48.2679	50.5111	55.5246
23	32.0069	35.1725	38.0756	41.6384	44.1813	49.7282	52.0002	57.0746
24	33.1962	36.4150	39.3641	42.9798	45.5585	51.1786	53.4788	58.6130
25	34.3816	37.6525	40.6465	44.3141	46.9279	52.6197	54.9475	60.1403
26	35.5632	38.8851	41.9232	45.6417	48.2899	54.0520	56.4069	61.6573
27	36.7412	40.1133	43.1945	46.9629	49.6449	55.4760	57.8576	63.1645
28	37.9159	41.3371	44.4608	48.2782	50.9934	56.8923	59.3000	64.6624
29	39.0875	42.5570	45.7223	49.5879	52.3356	58.3012	60.7346	66.1517
30	40.2560	43.7730	46.9792	50.8922	53.6720	59.7031	62.1619	67.6326
31	41.4217	44.9853	48.2319	52.1914	55.0027	61.0983	63.5820	69.1057
32	42.5847	46.1943	49.4804	53.4858	56.3281	62.4872	64.9955	70.5712
33	43.7452	47.3999	50.7251	54.7755	57.6484	63.8701	66.4025	72.0296
34	44.9032	48.6024	51.9660	56.0609	58.9639	65.2472	67.8035	73.4812
35	46.0588	49.8018	53.2033	57.3421	60.2748	66.6188	69.1986	74.9262
36	47.2122	50.9985	54.4373	58.6192	61.5812	67.9852	70.5881	76.3650
37	48.3634	52.1923	55.6680	59.8925	62.8833	69.3465	71.9722	77.7977
38	49.5126	53.3835	56.8955	61.1621	64.1814	70.7029	73.3512	79.2247
39	50.6598	54.5722	58.1201	62.4281	65.4756	72.0547	74.7253	80.6462
40	51.8051	55.7585	59.3417	63.6907	66.7660	73.4020	76.0946	82.0623
50	63.1671	67.5048	71.4202	76.1539	79.4900	86.6608	89.5605	95.9687
60	74.3970	79.0819	83.2977	88.3794	91.9517	99.6072	102.6948	109.5029
70	85.5270	90.5312	95.0232	100.4252	104.2149	112.3169	115.5776	122.7547
80	96.5782	101.8795	106.6286	112.3288	116.3211	124.8392	128.2613	135.7825
90	107.5650	113.1453	118.1359	124.1163	128.2989	137.2084	140.7823	148.6273
100	118.4980	124.3421	129.5612	135.8067	140.1695	149.4493	153.1670	161.3187

Table 10. Critical Values For The F Distribution

This table contains critical values F_{α,ν_1,ν_2} for the F distribution defined by $P(F \geq F_{\alpha,\nu_1,\nu_2}) = \alpha$.

$\alpha = .05$

$\nu_2 \backslash \nu_1$	1	2	3	4	5	6	7	8	9	10	15	20	30	40	60	120	∞
1	161.45	199.50	215.71	224.58	230.16	233.99	236.77	238.88	240.54	241.88	245.95	248.01	250.10	251.14	252.20	253.25	254.25
2	18.51	19.00	19.16	19.25	19.30	19.33	19.35	19.37	19.38	19.40	19.43	19.45	19.46	19.47	19.48	19.49	19.50
3	10.13	9.55	9.28	9.12	9.01	8.94	8.89	8.85	8.81	8.79	8.70	8.66	8.62	8.59	8.57	8.55	8.53
4	7.71	6.94	6.59	6.39	6.26	6.16	6.09	6.04	6.00	5.96	5.86	5.80	5.75	5.72	5.69	5.66	5.63
5	6.61	5.79	5.41	5.19	5.05	4.95	4.88	4.82	4.77	4.74	4.62	4.56	4.50	4.46	4.43	4.40	4.37
6	5.99	5.14	4.76	4.53	4.39	4.28	4.21	4.15	4.10	4.06	3.94	3.87	3.81	3.77	3.74	3.70	3.67
7	5.59	4.74	4.35	4.12	3.97	3.87	3.79	3.73	3.68	3.64	3.51	3.44	3.38	3.34	3.30	3.27	3.23
8	5.32	4.46	4.07	3.84	3.69	3.58	3.50	3.44	3.39	3.35	3.22	3.15	3.08	3.04	3.01	2.97	2.93
9	5.12	4.26	3.86	3.63	3.48	3.37	3.29	3.23	3.18	3.14	3.01	2.94	2.86	2.83	2.79	2.75	2.71
10	4.96	4.10	3.71	3.48	3.33	3.22	3.14	3.07	3.02	2.98	2.85	2.77	2.70	2.66	2.62	2.58	2.54
11	4.84	3.98	3.59	3.36	3.20	3.09	3.01	2.95	2.90	2.85	2.72	2.65	2.57	2.53	2.49	2.45	2.41
12	4.75	3.89	3.49	3.26	3.11	3.00	2.91	2.85	2.80	2.75	2.62	2.54	2.47	2.43	2.38	2.34	2.30
13	4.67	3.81	3.41	3.18	3.03	2.92	2.83	2.77	2.71	2.67	2.53	2.46	2.38	2.34	2.30	2.25	2.21
14	4.60	3.74	3.34	3.11	2.96	2.85	2.76	2.70	2.65	2.60	2.46	2.39	2.31	2.27	2.22	2.18	2.13
15	4.54	3.68	3.29	3.06	2.90	2.79	2.71	2.64	2.59	2.54	2.40	2.33	2.25	2.20	2.16	2.11	2.07
16	4.49	3.63	3.24	3.01	2.85	2.74	2.66	2.59	2.54	2.49	2.35	2.28	2.19	2.15	2.11	2.06	2.01
17	4.45	3.59	3.20	2.96	2.81	2.70	2.61	2.55	2.49	2.45	2.31	2.23	2.15	2.10	2.06	2.01	1.96
18	4.41	3.55	3.16	2.93	2.77	2.66	2.58	2.51	2.46	2.41	2.27	2.19	2.11	2.06	2.02	1.97	1.92
19	4.38	3.52	3.13	2.90	2.74	2.63	2.54	2.48	2.42	2.38	2.23	2.16	2.07	2.03	1.98	1.93	1.88
20	4.35	3.49	3.10	2.87	2.71	2.60	2.51	2.45	2.39	2.35	2.20	2.12	2.04	1.99	1.95	1.90	1.85
21	4.32	3.47	3.07	2.84	2.68	2.57	2.49	2.42	2.37	2.32	2.18	2.10	2.01	1.96	1.92	1.87	1.82
22	4.30	3.44	3.05	2.82	2.66	2.55	2.46	2.40	2.34	2.30	2.15	2.07	1.98	1.94	1.89	1.84	1.79
23	4.28	3.42	3.03	2.80	2.64	2.53	2.44	2.37	2.32	2.27	2.13	2.05	1.96	1.91	1.86	1.81	1.76
24	4.26	3.40	3.01	2.78	2.62	2.51	2.42	2.36	2.30	2.25	2.11	2.03	1.94	1.89	1.84	1.79	1.74
25	4.24	3.39	2.99	2.76	2.60	2.49	2.40	2.34	2.28	2.24	2.09	2.01	1.92	1.87	1.82	1.77	1.71
30	4.17	3.32	2.92	2.69	2.53	2.42	2.33	2.27	2.21	2.16	2.01	1.93	1.84	1.79	1.74	1.68	1.63
40	4.08	3.23	2.84	2.61	2.45	2.34	2.25	2.18	2.12	2.08	1.92	1.84	1.74	1.69	1.64	1.58	1.51
50	4.03	3.18	2.79	2.56	2.40	2.29	2.20	2.13	2.07	2.03	1.87	1.78	1.69	1.63	1.58	1.51	1.44
60	4.00	3.15	2.76	2.53	2.37	2.25	2.17	2.10	2.04	1.99	1.84	1.75	1.65	1.59	1.53	1.47	1.39
120	3.92	3.07	2.68	2.45	2.29	2.18	2.09	2.02	1.96	1.91	1.75	1.66	1.55	1.50	1.43	1.35	1.26
∞	3.85	3.00	2.61	2.38	2.22	2.10	2.01	1.94	1.88	1.84	1.67	1.58	1.46	1.40	1.32	1.23	1.00

Table 10. Critical Values For The F Distribution (Continued)

$\alpha = .01$

ν_1

ν_1	1	2	3	4	5	6	7	8	9	10	15	20	30	40	60	120	∞
2	98.50	99.00	99.17	99.25	99.30	99.33	99.36	99.37	99.39	99.40	99.43	99.45	99.47	99.47	99.48	99.49	99.50
3	34.12	30.82	29.46	28.71	28.24	27.91	27.67	27.49	27.35	27.23	26.87	26.69	26.50	26.41	26.32	26.22	26.13
4	21.20	18.00	16.69	15.98	15.52	15.21	14.98	14.80	14.66	14.55	14.20	14.02	13.84	13.75	13.65	13.56	13.47
5	16.26	13.27	12.06	11.39	10.97	10.67	10.46	10.29	10.16	10.05	9.72	9.55	9.38	9.29	9.20	9.11	9.03
6	13.75	10.92	9.78	9.15	8.75	8.47	8.26	8.10	7.98	7.87	7.56	7.40	7.23	7.14	7.06	6.97	6.89
7	12.25	9.55	8.45	7.85	7.46	7.19	6.99	6.84	6.72	6.62	6.31	6.16	5.99	5.91	5.82	5.74	5.65
8	11.26	8.65	7.59	7.01	6.63	6.37	6.18	6.03	5.91	5.81	5.52	5.36	5.20	5.12	5.03	4.95	4.86
9	10.56	8.02	6.99	6.42	6.06	5.80	5.61	5.47	5.35	5.26	4.96	4.81	4.65	4.57	4.48	4.40	4.32
10	10.04	7.56	6.55	5.99	5.64	5.39	5.20	5.06	4.94	4.85	4.56	4.41	4.25	4.17	4.08	4.00	3.91
11	9.65	7.21	6.22	5.67	5.32	5.07	4.89	4.74	4.63	4.54	4.25	4.10	3.94	3.86	3.78	3.69	3.61
12	9.33	6.93	5.95	5.41	5.06	4.82	4.64	4.50	4.39	4.30	4.01	3.86	3.70	3.62	3.54	3.45	3.37
13	9.07	6.70	5.74	5.21	4.86	4.62	4.44	4.30	4.19	4.10	3.82	3.66	3.51	3.43	3.34	3.25	3.17
14	8.86	6.51	5.56	5.04	4.69	4.46	4.28	4.14	4.03	3.94	3.66	3.51	3.35	3.27	3.18	3.09	3.01
15	8.68	6.36	5.42	4.89	4.56	4.32	4.14	4.00	3.89	3.80	3.52	3.37	3.21	3.13	3.05	2.96	2.87
16	8.53	6.23	5.29	4.77	4.44	4.20	4.03	3.89	3.78	3.69	3.41	3.26	3.10	3.02	2.93	2.84	2.76
17	8.40	6.11	5.19	4.67	4.34	4.10	3.93	3.79	3.68	3.59	3.31	3.16	3.00	2.92	2.83	2.75	2.66
18	8.29	6.01	5.09	4.58	4.25	4.01	3.84	3.71	3.60	3.51	3.23	3.08	2.92	2.84	2.75	2.66	2.57
19	8.18	5.93	5.01	4.50	4.17	3.94	3.77	3.63	3.52	3.43	3.15	3.00	2.84	2.76	2.67	2.58	2.50
20	8.10	5.85	4.94	4.43	4.10	3.87	3.70	3.56	3.46	3.37	3.09	2.94	2.78	2.69	2.61	2.52	2.43
21	8.02	5.78	4.87	4.37	4.04	3.81	3.64	3.51	3.40	3.31	3.03	2.88	2.72	2.64	2.55	2.46	2.37
22	7.95	5.72	4.82	4.31	3.99	3.76	3.59	3.45	3.35	3.26	2.98	2.83	2.67	2.58	2.50	2.40	2.31
23	7.88	5.66	4.76	4.26	3.94	3.71	3.54	3.41	3.30	3.21	2.93	2.78	2.62	2.54	2.45	2.35	2.26
24	7.82	5.61	4.72	4.22	3.90	3.67	3.50	3.36	3.26	3.17	2.89	2.74	2.58	2.49	2.40	2.31	2.22
25	7.77	5.57	4.68	4.18	3.85	3.63	3.46	3.32	3.22	3.13	2.85	2.70	2.54	2.45	2.36	2.27	2.18
30	7.56	5.39	4.51	4.02	3.70	3.47	3.30	3.17	3.07	2.98	2.70	2.55	2.39	2.30	2.21	2.11	2.01
40	7.31	5.18	4.31	3.83	3.51	3.29	3.12	2.99	2.89	2.80	2.52	2.37	2.20	2.11	2.02	1.92	1.81
50	7.17	5.06	4.20	3.72	3.41	3.19	3.02	2.89	2.78	2.70	2.42	2.27	2.10	2.01	1.91	1.80	1.69
60	7.08	4.98	4.13	3.65	3.34	3.12	2.95	2.82	2.72	2.63	2.35	2.20	2.03	1.94	1.84	1.73	1.61
120	6.85	4.79	3.95	3.48	3.17	2.96	2.79	2.66	2.56	2.47	2.19	2.03	1.86	1.76	1.66	1.53	1.39
∞	6.65	4.62	3.79	3.33	3.03	2.81	2.65	2.52	2.42	2.33	2.05	1.89	1.71	1.60	1.48	1.34	1.00

Table 10. Critical Values For The F Distribution (Continued)

α = .001

ν_1	1	2	3	4	5	6	7	8	9	10	15	20	30	40	60	120	∞
2	998.50	999.00	999.17	999.25	999.30	999.33	999.36	999.37	999.39	999.40	999.43	999.45	999.47	999.47	999.48	999.49	999.50
3	167.03	148.50	141.11	137.10	134.58	132.85	131.58	130.62	129.86	129.25	127.37	126.42	125.45	124.96	124.47	123.97	123.50
4	74.14	61.25	56.18	53.44	51.71	50.53	49.66	49.00	48.47	48.05	46.76	46.10	45.43	45.09	44.75	44.40	44.07
5	47.18	37.12	33.20	31.09	29.75	28.83	28.16	27.65	27.24	26.92	25.91	25.39	24.87	24.60	24.33	24.06	23.80
6	35.51	27.00	23.70	21.92	20.80	20.03	19.46	19.03	18.69	18.41	17.56	17.12	16.67	16.44	16.21	15.98	15.76
7	29.25	21.69	18.77	17.20	16.21	15.52	15.02	14.63	14.33	14.08	13.32	12.93	12.53	12.33	12.12	11.91	11.71
8	25.41	18.49	15.83	14.39	13.48	12.86	12.40	12.05	11.77	11.54	10.84	10.48	10.11	9.92	9.73	9.53	9.35
9	22.86	16.39	13.90	12.56	11.71	11.13	10.70	10.37	10.11	9.89	9.24	8.90	8.55	8.37	8.19	8.00	7.82
10	21.04	14.91	12.55	11.28	10.48	9.93	9.52	9.20	8.96	8.75	8.13	7.80	7.47	7.30	7.12	6.94	6.77
11	19.69	13.81	11.56	10.35	9.58	9.05	8.66	8.35	8.12	7.92	7.32	7.01	6.68	6.52	6.35	6.18	6.01
12	18.64	12.97	10.80	9.63	8.89	8.38	8.00	7.71	7.48	7.29	6.71	6.40	6.09	5.93	5.76	5.59	5.43
13	17.82	12.31	10.21	9.07	8.35	7.86	7.49	7.21	6.98	6.80	6.23	5.93	5.63	5.47	5.30	5.14	4.98
14	17.14	11.78	9.73	8.62	7.92	7.44	7.08	6.80	6.58	6.40	5.85	5.56	5.25	5.10	4.94	4.77	4.61
15	16.59	11.34	9.34	8.25	7.57	7.09	6.74	6.47	6.26	6.08	5.54	5.25	4.95	4.80	4.64	4.47	4.32
16	16.12	10.97	9.01	7.94	7.27	6.80	6.46	6.19	5.98	5.81	5.27	4.99	4.70	4.54	4.39	4.23	4.07
17	15.72	10.66	8.73	7.68	7.02	6.56	6.22	5.96	5.75	5.58	5.05	4.78	4.48	4.33	4.18	4.02	3.86
18	15.38	10.39	8.49	7.46	6.81	6.35	6.02	5.76	5.56	5.39	4.87	4.59	4.30	4.15	4.00	3.84	3.68
19	15.08	10.16	8.28	7.27	6.62	6.18	5.85	5.59	5.39	5.22	4.70	4.43	4.14	3.99	3.84	3.68	3.52
20	14.82	9.95	8.10	7.10	6.46	6.02	5.69	5.44	5.24	5.08	4.56	4.29	4.00	3.86	3.70	3.54	3.39
21	14.59	9.77	7.94	6.95	6.32	5.88	5.56	5.31	5.11	4.95	4.44	4.17	3.88	3.74	3.58	3.42	3.27
22	14.38	9.61	7.80	6.81	6.19	5.76	5.44	5.19	4.99	4.83	4.33	4.06	3.78	3.63	3.48	3.32	3.16
23	14.20	9.47	7.67	6.70	6.08	5.65	5.33	5.09	4.89	4.73	4.23	3.96	3.68	3.53	3.38	3.22	3.07
24	14.03	9.34	7.55	6.59	5.98	5.55	5.23	4.99	4.80	4.64	4.14	3.87	3.59	3.45	3.29	3.14	2.98
25	13.88	9.22	7.45	6.49	5.89	5.46	5.15	4.91	4.71	4.56	4.06	3.79	3.52	3.37	3.22	3.06	2.90
30	13.29	8.77	7.05	6.12	5.53	5.12	4.82	4.58	4.39	4.24	3.75	3.49	3.22	3.07	2.92	2.76	2.60
40	12.61	8.25	6.59	5.70	5.13	4.73	4.44	4.21	4.02	3.87	3.40	3.14	2.87	2.73	2.57	2.41	2.24
50	12.22	7.96	6.34	5.46	4.90	4.51	4.22	4.00	3.82	3.67	3.20	2.95	2.68	2.53	2.38	2.21	2.04
60	11.97	7.77	6.17	5.31	4.76	4.37	4.09	3.86	3.69	3.54	3.08	2.83	2.55	2.41	2.25	2.08	1.90
120	11.38	7.32	5.78	4.95	4.42	4.04	3.77	3.55	3.38	3.24	2.78	2.53	2.26	2.11	1.95	1.77	1.56
∞	10.86	6.93	5.44	4.64	4.12	3.76	3.49	3.28	3.11	2.97	2.53	2.28	2.01	1.85	1.68	1.47	1.00

ν_1

Table 11. The Incomplete Gamma Function

This table contains values of $F(x; \alpha) = \int_0^x \frac{1}{\Gamma(\alpha)} y^{\alpha-1} e^{-y}\, dy.$

					α					
x	0.5	1.0	1.5	2.0	2.5	3.0	3.5	4.0	4.5	5.0
1	.8427	.6321	.4276	.2642	.1509	.0803	.0402	.0190	.0085	.0037
2	.9545	.8647	.7385	.5940	.4506	.3233	.2202	.1429	.0886	.0527
3	.9857	.9502	.8884	.8009	.6938	.5768	.4603	.3528	.2601	.1847
4	.9953	.9817	.9540	.9084	.8438	.7619	.6674	.5665	.4659	.3712
5	.9984	.9933	.9814	.9596	.9248	.8753	.8114	.7350	.6495	.5595
6	.9995	.9975	.9926	.9826	.9652	.9380	.8994	.8488	.7867	.7149
7	.9998	.9991	.9971	.9927	.9844	.9704	.9488	.9182	.8777	.8270
8	.9999	.9997	.9989	.9970	.9932	.9862	.9749	.9576	.9331	.9004
9	1.0000	.9999	.9996	.9988	.9971	.9938	.9880	.9788	.9648	.9450
10		1.0000	.9998	.9995	.9988	.9972	.9944	.9897	.9821	.9707
11			.9999	.9998	.9995	.9988	.9975	.9951	.9911	.9849
12			1.0000	.9999	.9998	.9995	.9989	.9977	.9957	.9924
13				1.0000	.9999	.9998	.9995	.9989	.9980	.9963
14					1.0000	.9999	.9998	.9995	.9990	.9982
15						1.0000	.9999	.9998	.9996	.9991
16							1.0000	.9999	.9998	.9996
17								1.0000	.9999	.9998
18									1.0000	.9999
19										1.0000

					α					
x	5.5	6.0	6.5	7.0	7.5	8.0	8.5	9.0	9.5	10.0
1	.0015	.0006	.0002	.0001	.0000	.0000	.0000	.0000	.0000	
2	.0301	.0166	.0088	.0045	.0023	.0011	.0005	.0002	.0001	.0000
3	.1266	.0839	.0538	.0335	.0203	.0119	.0068	.0038	.0021	.0011
4	.2867	.2149	.1564	.1107	.0762	.0511	.0335	.0214	.0133	.0081
5	.4696	.3840	.3061	.2378	.1803	.1334	.0964	.0681	.0471	.0318
6	.6364	.5543	.4724	.3937	.3210	.2560	.1999	.1528	.1144	.0839
7	.7670	.6993	.6262	.5503	.4745	.4013	.3329	.2709	.2163	.1695
8	.8589	.8088	.7509	.6866	.6179	.5470	.4762	.4075	.3427	.2834
9	.9184	.8843	.8425	.7932	.7373	.6761	.6112	.5443	.4776	.4126
10	.9547	.9329	.9048	.8699	.8281	.7798	.7258	.6672	.6054	.5421
11	.9756	.9625	.9446	.9214	.8922	.8568	.8153	.7680	.7157	.6595
12	.9873	.9797	.9689	.9542	.9349	.9105	.8806	.8450	.8038	.7576
13	.9935	.9893	.9830	.9741	.9620	.9460	.9255	.9002	.8698	.8342
14	.9968	.9945	.9910	.9858	.9784	.9684	.9551	.9379	.9166	.8906
15	.9984	.9972	.9953	.9924	.9881	.9820	.9737	.9626	.9482	.9301
16	.9992	.9986	.9976	.9960	.9936	.9900	.9850	.9780	.9687	.9567
17	.9996	.9993	.9988	.9979	.9966	.9946	.9916	.9874	.9816	.9739
18	.9998	.9997	.9994	.9990	.9982	.9971	.9954	.9929	.9894	.9846
19	.9999	.9998	.9997	.9995	.9991	.9985	.9975	.9961	.9941	.9911
20	1.0000	.9999	.9999	.9997	.9995	.9992	.9987	.9979	.9967	.9950
21		1.0000	.9999	.9999	.9998	.9996	.9993	.9989	.9982	.9972
22			1.0000	.9999	.9999	.9998	.9997	.9994	.9991	.9985
23				1.0000	.9999	.9999	.9998	.9997	.9995	.9992
24					1.0000	1.0000	.9999	.9998	.9997	.9996

63

Table 12. Critical Values For The Studentized Range Distribution

This table contains critical values $Q_{\alpha,\nu}$ for the Studentized Range distribution defined by $P(Q \geq Q_{\alpha,k,\nu}) = \alpha$, k is the number of degrees of freedom in the numerator (the number of treatment groups) and ν is the number of degrees of freedom in the denominator (s^2).

$\alpha = .05$

k

ν	2	3	4	5	6	7	8	9	10	11	12	13	14	15	16	17	18	19	20
1	17.97	26.98	32.82	37.08	40.41	43.12	45.40	47.36	49.07	50.59	51.96	53.20	54.33	55.36	56.32	57.22	58.04	58.83	59.56
2	6.085	8.331	9.798	10.88	11.74	12.44	13.03	13.54	13.99	14.39	14.75	15.08	15.38	15.65	15.91	16.14	16.37	16.57	16.77
3	4.501	5.910	6.825	7.502	8.037	8.478	8.853	9.177	9.462	9.717	9.946	10.15	10.35	10.53	10.69	10.84	10.98	11.11	11.24
4	3.927	5.040	5.757	6.287	6.707	7.053	7.347	7.602	7.826	8.027	8.208	8.373	8.525	8.664	8.794	8.914	9.028	9.134	9.233
5	3.635	4.602	5.218	5.673	6.033	6.330	6.582	6.802	6.995	7.168	7.324	7.466	7.596	7.717	7.828	7.932	8.030	8.122	8.208
6	3.461	4.339	4.896	5.305	5.628	5.895	6.122	6.319	6.493	6.649	6.789	6.917	7.034	7.143	7.244	7.338	7.426	7.508	7.587
7	3.344	4.165	4.681	5.060	5.359	5.606	5.815	5.998	6.158	6.302	6.431	6.550	6.658	6.759	6.852	6.939	7.020	7.097	7.170
8	3.261	4.041	4.529	4.886	5.167	5.399	5.597	5.767	5.918	6.054	6.175	6.287	6.389	6.483	6.571	6.653	6.729	6.802	6.870
9	3.199	3.949	4.415	4.756	5.024	5.244	5.432	5.595	5.739	5.867	5.983	6.089	6.186	6.276	6.359	6.437	6.510	6.579	6.644
10	3.151	3.877	4.327	4.654	4.912	5.124	5.305	5.461	5.599	5.722	5.833	5.935	6.028	6.114	6.194	6.269	6.339	6.405	6.467
11	3.113	3.820	4.256	4.574	4.823	5.028	5.202	5.353	5.487	5.605	5.713	5.811	5.901	5.984	6.062	6.134	6.202	6.265	6.326
12	3.082	3.773	4.199	4.508	4.751	4.950	5.119	5.265	5.395	5.511	5.615	5.710	5.798	5.878	5.953	6.023	6.089	6.151	6.209
13	3.055	3.735	4.151	4.453	4.690	4.885	5.049	5.192	5.318	5.431	5.533	5.625	5.711	5.789	5.862	5.931	5.995	6.055	6.112
14	3.033	3.702	4.111	4.407	4.639	4.829	4.990	5.131	5.254	5.364	5.463	5.554	5.637	5.714	5.786	5.852	5.915	5.974	6.029
15	3.014	3.674	4.076	4.367	4.595	4.782	4.940	5.077	5.198	5.306	5.404	5.493	5.574	5.649	5.720	5.785	5.846	5.904	5.958
16	2.998	3.649	4.046	4.333	4.557	4.741	4.897	5.031	5.150	5.256	5.352	5.439	5.520	5.593	5.662	5.727	5.786	5.843	5.897
17	2.984	3.628	4.020	4.303	4.524	4.705	4.858	4.991	5.108	5.212	5.307	5.392	5.471	5.544	5.612	5.675	5.734	5.790	5.842
18	2.971	3.609	3.997	4.277	4.495	4.673	4.824	4.956	5.071	5.174	5.267	5.352	5.429	5.501	5.568	5.630	5.688	5.743	5.794
19	2.960	3.593	3.977	4.253	4.469	4.645	4.794	4.924	5.038	5.140	5.231	5.315	5.391	5.462	5.528	5.589	5.647	5.701	5.752
20	2.950	3.578	3.958	4.232	4.445	4.620	4.768	4.896	5.008	5.108	5.199	5.282	5.357	5.427	5.493	5.553	5.610	5.663	5.714
24	2.919	3.532	3.901	4.166	4.373	4.541	4.684	4.807	4.915	5.012	5.099	5.179	5.251	5.319	5.381	5.439	5.494	5.545	5.594
30	2.888	3.486	3.845	4.102	4.302	4.464	4.602	4.720	4.824	4.917	5.001	5.077	5.147	5.211	5.271	5.327	5.379	5.429	5.475
40	2.858	3.442	3.791	4.039	4.232	4.389	4.521	4.635	4.735	4.824	4.904	4.977	5.044	5.106	5.163	5.216	5.266	5.313	5.358
60	2.829	3.399	3.737	3.977	4.163	4.314	4.441	4.550	4.646	4.732	4.808	4.878	4.942	5.001	5.056	5.107	5.154	5.199	5.241
120	2.800	3.356	3.685	3.917	4.096	4.241	4.363	4.468	4.560	4.641	4.714	4.781	4.842	4.898	4.950	4.998	5.044	5.086	5.126
∞	2.772	3.314	3.633	3.858	4.030	4.170	4.286	4.387	4.474	4.552	4.622	4.685	4.743	4.796	4.845	4.891	4.934	4.974	5.012

Table 12. Critical Values For The Studentized Range Distribution (Continued)

α = .01

k

ν	2	3	4	5	6	7	8	9	10	11	12	13	14	15	16	17	18	19	20
1	90.03	135.0	164.3	185.6	202.2	215.8	227.2	237.0	245.6	253.2	260.0	266.2	271.8	277.0	281.8	286.3	290.4	294.3	298.0
2	14.04	19.02	22.29	24.72	26.63	28.20	29.53	30.68	31.69	32.59	33.40	34.13	34.81	35.43	36.00	36.53	37.03	37.50	37.95
3	8.261	10.62	12.17	13.33	14.24	15.00	15.64	16.20	16.69	17.13	17.53	17.89	18.22	18.52	18.81	19.07	19.32	19.55	19.77
4	6.512	8.120	9.173	9.958	10.58	11.10	11.55	11.93	12.27	12.57	12.84	13.09	13.32	13.53	13.73	13.91	14.08	14.24	14.40
5	5.702	6.976	7.804	8.421	8.913	9.321	9.669	9.972	10.24	10.48	10.70	10.89	11.08	11.24	11.40	11.55	11.68	11.81	11.93
6	5.243	6.331	7.033	7.556	7.973	8.318	8.613	8.869	9.097	9.301	9.485	9.653	9.808	9.951	10.08	10.21	10.32	10.43	10.54
7	4.949	5.919	6.543	7.005	7.373	7.679	7.939	8.166	8.368	8.548	8.711	8.860	8.997	9.124	9.242	9.353	9.456	9.554	9.646
8	4.746	5.635	6.204	6.625	6.960	7.237	7.474	7.681	7.863	8.027	8.176	8.312	8.436	8.552	8.659	8.760	8.854	8.943	9.027
9	4.596	5.428	5.957	6.348	6.658	6.915	7.134	7.325	7.495	7.647	7.784	7.910	8.025	8.132	8.232	8.325	8.412	8.495	8.573
10	4.482	5.270	5.769	6.136	6.428	6.669	6.875	7.055	7.213	7.356	7.485	7.603	7.712	7.812	7.906	7.993	8.076	8.153	8.226
11	4.392	5.146	5.621	5.970	6.247	6.476	6.672	6.842	6.992	7.128	7.250	7.362	7.465	7.560	7.649	7.732	7.809	7.883	7.952
12	4.320	5.046	5.502	5.836	6.101	6.321	6.507	6.670	6.814	6.943	7.060	7.167	7.265	7.356	7.441	7.520	7.594	7.665	7.731
13	4.260	4.964	5.404	5.727	5.981	6.192	6.372	6.528	6.667	6.791	6.903	7.006	7.101	7.188	7.269	7.345	7.417	7.485	7.548
14	4.210	4.895	5.322	5.634	5.881	6.085	6.258	6.409	6.543	6.664	6.772	6.871	6.962	7.047	7.126	7.199	7.268	7.333	7.395
15	4.168	4.836	5.252	5.556	5.796	5.994	6.162	6.309	6.439	6.555	6.660	6.757	6.845	6.927	7.003	7.074	7.142	7.204	7.264
16	4.131	4.786	5.192	5.489	5.722	5.915	6.079	6.222	6.349	6.462	6.564	6.658	6.744	6.823	6.898	6.967	7.032	7.093	7.152
17	4.099	4.742	5.140	5.430	5.659	5.847	6.007	6.147	6.270	6.381	6.480	6.572	6.656	6.734	6.806	6.873	6.937	6.997	7.053
18	4.071	4.703	5.094	5.379	5.603	5.788	5.944	6.081	6.201	6.310	6.407	6.497	6.579	6.655	6.725	6.792	6.854	6.912	6.968
19	4.046	4.670	5.054	5.334	5.554	5.735	5.889	6.022	6.141	6.247	6.342	6.430	6.510	6.585	6.654	6.719	6.780	6.837	6.891
20	4.024	4.639	5.018	5.294	5.510	5.688	5.839	5.970	6.087	6.191	6.285	6.371	6.450	6.523	6.591	6.654	6.714	6.771	6.823
24	3.956	4.546	4.907	5.168	5.374	5.542	5.685	5.809	5.919	6.017	6.106	6.186	6.261	6.330	6.394	6.453	6.510	6.563	6.612
30	3.889	4.455	4.799	5.048	5.242	5.401	5.536	5.653	5.756	5.849	5.932	6.008	6.078	6.143	6.203	6.259	6.311	6.361	6.407
40	3.825	4.367	4.696	4.931	5.114	5.265	5.392	5.502	5.599	5.686	5.764	5.835	5.900	5.961	6.017	6.069	6.119	6.165	6.209
60	3.762	4.282	4.595	4.818	4.991	5.133	5.253	5.356	5.447	5.528	5.601	5.667	5.728	5.785	5.837	5.886	5.931	5.974	6.015
120	3.702	4.200	4.497	4.709	4.872	5.005	5.118	5.214	5.299	5.375	5.443	5.505	5.562	5.614	5.662	5.708	5.750	5.790	5.827
∞	3.643	4.120	4.403	4.603	4.757	4.882	4.987	5.078	5.157	5.227	5.290	5.348	5.400	5.448	5.493	5.535	5.574	5.611	5.645

Table 12. Critical Values For The Studentized Range Distribution (Continued)

α = .001

ν	k																		
	2	3	4	5	6	7	8	9	10	11	12	13	14	15	16	17	18	19	20
1	900.3	1351.	1643.	1856.	2022.	2158.	2272.	2370.	2455.	2532.	2600.	2662.	2718.	2770.	2818.	2863.	2904.	2943.	2980.
2	44.69	60.42	70.77	78.43	84.49	89.46	93.67	97.30	100.5	103.3	105.9	108.2	110.4	112.3	114.2	115.9	117.4	118.9	120.3
3	18.28	23.32	26.65	29.13	31.11	32.74	34.12	35.33	36.39	37.34	38.20	38.98	39.69	40.35	40.97	41.54	42.07	42.58	43.05
4	12.18	14.99	16.84	18.23	19.34	20.26	21.04	21.73	22.33	22.87	23.36	23.81	24.21	24.59	24.94	25.27	25.58	25.87	26.14
5	9.714	11.67	12.96	13.93	14.71	15.35	15.90	16.38	16.81	17.18	17.53	17.85	18.13	18.41	18.66	18.89	19.10	19.31	19.51
6	8.427	9.960	10.97	11.72	12.32	12.83	13.26	13.63	13.97	14.27	14.54	14.79	15.01	15.22	15.42	15.60	15.78	15.94	16.09
7	7.648	8.930	9.768	10.40	10.90	11.32	11.68	11.99	12.27	12.52	12.74	12.95	13.14	13.32	13.48	13.64	13.78	13.92	14.04
8	7.130	8.250	8.978	9.522	9.958	10.32	10.64	10.91	11.15	11.36	11.56	11.74	11.91	12.06	12.21	12.34	12.47	12.59	12.70
9	6.762	7.768	8.419	8.906	9.295	9.619	9.897	10.14	10.36	10.55	10.73	10.89	11.03	11.18	11.30	11.42	11.54	11.64	11.75
10	6.487	7.411	8.006	8.450	8.804	9.099	9.352	9.573	9.769	9.946	10.11	10.25	10.39	10.52	10.64	10.75	10.85	10.95	11.03
11	6.275	7.136	7.687	8.098	8.426	8.699	8.933	9.138	9.319	9.482	9.630	9.766	9.892	10.01	10.12	10.22	10.31	10.41	10.49
12	6.106	6.917	7.436	7.821	8.127	8.383	8.601	8.793	8.962	9.115	9.254	9.381	9.498	9.606	9.707	9.802	9.891	9.975	10.06
13	5.970	6.740	7.231	7.595	7.885	8.126	8.333	8.513	8.673	8.817	8.948	9.068	9.178	9.281	9.376	9.466	9.550	9.629	9.704
14	5.856	6.594	7.062	7.409	7.685	7.915	8.110	8.282	8.434	8.571	8.696	8.809	8.914	9.012	9.103	9.188	9.267	9.343	9.414
15	5.760	6.470	6.920	7.252	7.517	7.736	7.925	8.088	8.234	8.365	8.483	8.592	8.693	8.786	8.872	8.954	9.030	9.102	9.170
16	5.678	6.365	6.799	7.119	7.374	7.585	7.766	7.923	8.063	8.189	8.303	8.407	8.504	8.593	8.676	8.755	8.828	8.897	8.963
17	5.608	6.275	6.695	7.005	7.250	7.454	7.629	7.781	7.916	8.037	8.148	8.248	8.342	8.427	8.508	8.583	8.654	8.720	8.784
18	5.546	6.196	6.604	6.905	7.143	7.341	7.510	7.657	7.788	7.906	8.012	8.110	8.199	8.283	8.361	8.434	8.502	8.567	8.628
19	5.492	6.127	6.525	6.817	7.049	7.242	7.405	7.549	7.676	7.790	7.893	7.988	8.075	8.156	8.232	8.303	8.369	8.432	8.491
20	5.444	6.065	6.454	6.740	6.966	7.154	7.313	7.453	7.577	7.688	7.788	7.880	7.966	8.044	8.118	8.186	8.251	8.312	8.370
24	5.297	5.877	6.238	6.503	6.712	6.884	7.031	7.159	7.272	7.374	7.467	7.551	7.629	7.701	7.768	7.831	7.890	7.946	7.999
30	5.156	5.698	6.033	6.278	6.470	6.628	6.763	6.880	6.984	7.077	7.162	7.239	7.310	7.375	7.437	7.494	7.548	7.599	7.647
40	5.022	5.528	5.838	6.063	6.240	6.386	6.509	6.616	6.711	6.796	6.872	6.942	7.007	7.067	7.122	7.174	7.223	7.269	7.312
60	4.894	5.365	5.653	5.860	6.022	6.155	6.268	6.366	6.451	6.528	6.598	6.661	6.720	6.774	6.824	6.871	6.914	6.956	6.995
120	4.771	5.211	5.476	5.667	5.815	5.937	6.039	6.128	6.206	6.276	6.339	6.396	6.448	6.496	6.542	6.583	6.623	6.660	6.695
∞	4.654	5.063	5.309	5.484	5.619	5.730	5.823	5.903	5.973	6.036	6.092	6.144	6.191	6.234	6.274	6.312	6.347	6.380	6.411

Table 13. Least Signficant Studentized Ranges For Duncan's Test

This table contains critical values or least significant studentized ranges, $r_{\alpha,p,\nu}$, for Duncan's Multiple Range Test where α is the significance level, p is the number of successive values from an ordered list of k means of equal sample sizes ($p = 2, 3, \ldots, k$), and ν is the degrees of freedom for the independent estimate s^2.

$\alpha = .05$

ν	2	3	4	5	6	7	8	9	10	11	12	13	14	15	16	17	18	19	20
1	17.97	17.97	17.97	17.97	17.97	17.97	17.97	17.97	17.97	17.97	17.97	17.97	17.97	17.97	17.97	17.97	17.97	17.97	17.97
2	6.085	6.085	6.085	6.085	6.085	6.085	6.085	6.085	6.085	6.085	6.085	6.085	6.085	6.085	6.085	6.085	6.085	6.085	6.085
3	4.501	4.516	4.516	4.516	4.516	4.516	4.516	4.516	4.516	4.516	4.516	4.516	4.516	4.516	4.516	4.516	4.516	4.516	4.516
4	3.927	4.013	4.033	4.033	4.033	4.033	4.033	4.033	4.033	4.033	4.033	4.033	4.033	4.033	4.033	4.033	4.033	4.033	4.033
5	3.635	3.749	3.797	3.814	3.814	3.814	3.814	3.814	3.814	3.814	3.814	3.814	3.814	3.814	3.814	3.814	3.814	3.814	3.814
6	3.461	3.587	3.649	3.680	3.694	3.697	3.697	3.697	3.697	3.697	3.697	3.697	3.697	3.697	3.697	3.697	3.697	3.697	3.697
7	3.344	3.477	3.548	3.588	3.611	3.622	3.626	3.626	3.626	3.626	3.626	3.626	3.626	3.626	3.626	3.626	3.626	3.626	3.626
8	3.261	3.399	3.475	3.521	3.549	3.566	3.575	3.579	3.579	3.579	3.579	3.579	3.579	3.579	3.579	3.579	3.579	3.579	3.579
9	3.199	3.339	3.420	3.470	3.502	3.523	3.536	3.544	3.547	3.547	3.547	3.547	3.547	3.547	3.547	3.547	3.547	3.547	3.547
10	3.151	3.293	3.376	3.430	3.465	3.489	3.505	3.516	3.522	3.525	3.526	3.526	3.526	3.526	3.526	3.526	3.526	3.526	3.526
11	3.113	3.256	3.342	3.397	3.435	3.462	3.480	3.493	3.501	3.506	3.509	3.510	3.510	3.510	3.510	3.510	3.510	3.510	3.510
12	3.082	3.225	3.313	3.370	3.410	3.439	3.459	3.474	3.484	3.491	3.496	3.498	3.499	3.499	3.499	3.499	3.499	3.499	3.499
13	3.055	3.200	3.289	3.348	3.389	3.419	3.442	3.458	3.470	3.478	3.484	3.488	3.490	3.490	3.490	3.490	3.490	3.490	3.490
14	3.033	3.178	3.268	3.329	3.372	3.403	3.426	3.444	3.457	3.467	3.474	3.479	3.482	3.484	3.484	3.485	3.485	3.485	3.485
15	3.014	3.160	3.250	3.312	3.356	3.389	3.413	3.432	3.446	3.457	3.465	3.471	3.476	3.478	3.480	3.481	3.481	3.481	3.481
16	2.998	3.144	3.235	3.298	3.343	3.376	3.402	3.422	3.437	3.449	3.458	3.465	3.470	3.473	3.477	3.478	3.478	3.478	3.478
17	2.984	3.130	3.222	3.285	3.331	3.366	3.392	3.412	3.429	3.441	3.451	3.459	3.465	3.469	3.473	3.475	3.476	3.476	3.476
18	2.971	3.118	3.210	3.274	3.321	3.356	3.383	3.405	3.421	3.435	3.445	3.454	3.460	3.465	3.470	3.472	3.474	3.474	3.474
19	2.960	3.107	3.199	3.264	3.311	3.347	3.375	3.397	3.415	3.429	3.440	3.449	3.456	3.462	3.467	3.470	3.472	3.473	3.474
20	2.950	3.097	3.190	3.255	3.303	3.339	3.368	3.391	3.409	3.424	3.436	3.445	3.453	3.459	3.464	3.467	3.470	3.472	3.473
24	2.919	3.066	3.160	3.226	3.276	3.315	3.345	3.370	3.390	3.406	3.420	3.432	3.441	3.449	3.456	3.461	3.465	3.469	3.471
30	2.888	3.035	3.131	3.199	3.250	3.290	3.322	3.349	3.371	3.389	3.405	3.418	3.430	3.439	3.447	3.454	3.460	3.466	3.470
40	2.858	3.006	3.102	3.171	3.224	3.266	3.300	3.328	3.352	3.373	3.390	3.405	3.418	3.429	3.439	3.448	3.456	3.463	3.469
60	2.829	2.976	3.073	3.143	3.198	3.241	3.277	3.307	3.333	3.355	3.374	3.391	3.406	3.419	3.431	3.442	3.451	3.460	3.467
120	2.800	2.947	3.045	3.116	3.172	3.217	3.254	3.287	3.314	3.337	3.359	3.377	3.394	3.409	3.423	3.435	3.446	3.457	3.466
∞	2.772	2.918	3.017	3.089	3.146	3.193	3.232	3.265	3.294	3.320	3.343	3.363	3.382	3.399	3.414	3.428	3.442	3.454	3.466

Table 13. Least Significant Studentized Ranges For Duncan's Test (Continued)

α = .01

ν	2	3	4	5	6	7	8	9	10	11	12	13	14	15	16	17	18	19	20
1	90.03	90.03	90.03	90.03	90.03	90.03	90.03	90.03	90.03	90.03	90.03	90.03	90.03	90.03	90.03	90.03	90.03	90.03	90.03
2	14.04	14.04	14.04	14.04	14.04	14.04	14.04	14.04	14.04	14.04	14.04	14.04	14.04	14.04	14.04	14.04	14.04	14.04	14.04
3	8.261	8.321	8.321	8.321	8.321	8.321	8.321	8.321	8.321	8.321	8.321	8.321	8.321	8.321	8.321	8.321	8.321	8.321	8.321
4	6.512	6.677	6.740	6.756	6.756	6.756	6.756	6.756	6.756	6.756	6.756	6.756	6.756	6.756	6.756	6.756	6.756	6.756	6.756
5	5.702	5.893	5.989	6.040	6.065	6.074	6.074	6.074	6.074	6.074	6.074	6.074	6.074	6.074	6.074	6.074	6.074	6.074	6.074
6	5.243	5.439	5.549	5.614	5.655	5.680	5.694	5.701	5.703	5.703	5.703	5.703	5.703	5.703	5.703	5.703	5.703	5.703	5.703
7	4.949	5.145	5.260	5.334	5.383	5.416	5.439	5.454	5.464	5.470	5.472	5.472	5.472	5.472	5.472	5.472	5.472	5.472	5.472
8	4.746	4.939	5.057	5.135	5.189	5.227	5.256	5.276	5.291	5.302	5.309	5.314	5.316	5.317	5.317	5.317	5.317	5.317	5.317
9	4.596	4.787	4.906	4.986	5.043	5.086	5.118	5.142	5.160	5.174	5.185	5.193	5.199	5.203	5.205	5.206	5.206	5.206	5.206
10	4.482	4.671	4.790	4.871	4.931	4.975	5.010	5.037	5.058	5.074	5.088	5.098	5.106	5.112	5.117	5.120	5.122	5.124	5.124
11	4.392	4.579	4.697	4.780	4.841	4.887	4.924	4.952	4.975	4.994	5.009	5.021	5.031	5.039	5.045	5.050	5.054	5.057	5.059
12	4.320	4.504	4.622	4.706	4.767	4.815	4.852	4.883	4.907	4.927	4.944	4.958	4.969	4.978	4.986	4.993	4.998	5.002	5.006
13	4.260	4.442	4.560	4.644	4.706	4.755	4.793	4.824	4.850	4.872	4.889	4.904	4.917	4.928	4.937	4.944	4.950	4.956	4.960
14	4.210	4.391	4.508	4.591	4.654	4.704	4.743	4.775	4.802	4.824	4.843	4.859	4.872	4.884	4.894	4.902	4.910	4.916	4.921
15	4.168	4.347	4.463	4.547	4.610	4.660	4.700	4.733	4.760	4.783	4.803	4.820	4.834	4.846	4.857	4.866	4.874	4.881	4.887
16	4.131	4.309	4.425	4.509	4.572	4.622	4.663	4.696	4.724	4.748	4.768	4.786	4.800	4.813	4.825	4.835	4.844	4.851	4.858
17	4.099	4.275	4.391	4.475	4.539	4.589	4.630	4.664	4.693	4.717	4.738	4.756	4.771	4.785	4.797	4.807	4.816	4.824	4.832
18	4.071	4.246	4.362	4.445	4.509	4.560	4.601	4.635	4.664	4.689	4.711	4.729	4.745	4.759	4.772	4.783	4.792	4.801	4.808
19	4.046	4.220	4.335	4.419	4.483	4.534	4.575	4.610	4.639	4.665	4.686	4.705	4.722	4.736	4.749	4.761	4.771	4.780	4.788
20	4.024	4.197	4.312	4.395	4.459	4.510	4.552	4.587	4.617	4.642	4.664	4.684	4.701	4.716	4.729	4.741	4.751	4.761	4.769
24	3.956	4.126	4.239	4.322	4.386	4.437	4.480	4.516	4.546	4.573	4.596	4.616	4.634	4.651	4.665	4.678	4.690	4.700	4.710
30	3.889	4.056	4.168	4.250	4.314	4.366	4.409	4.445	4.477	4.504	4.528	4.550	4.569	4.586	4.601	4.615	4.628	4.640	4.650
40	3.825	3.988	4.098	4.180	4.244	4.296	4.339	4.376	4.408	4.436	4.461	4.483	4.503	4.521	4.537	4.553	4.566	4.579	4.591
60	3.762	3.922	4.031	4.111	4.174	4.226	4.270	4.307	4.340	4.368	4.394	4.417	4.438	4.456	4.474	4.490	4.504	4.518	4.530
120	3.702	3.858	3.965	4.044	4.107	4.158	4.202	4.239	4.272	4.301	4.327	4.351	4.372	4.392	4.410	4.426	4.442	4.456	4.469
∞	3.643	3.796	3.900	3.978	4.040	4.091	4.135	4.172	4.205	4.235	4.261	4.285	4.307	4.327	4.345	4.363	4.379	4.394	4.408

Table 13. Least Signficant Studentized Ranges For Duncan's Test (Continued)

α = .001

ν										p									
	2	3	4	5	6	7	8	9	10	11	12	13	14	15	16	17	18	19	20
1	900.3	900.3	900.3	900.3	900.3	900.3	900.3	900.3	900.3	900.3	900.3	900.3	900.3	900.3	900.3	900.3	900.3	900.3	900.3
2	44.69	44.69	44.69	44.69	44.69	44.69	44.69	44.69	44.69	44.69	44.69	44.69	44.69	44.69	44.69	44.69	44.69	44.69	44.69
3	18.28	18.45	18.45	18.45	18.45	18.45	18.45	18.45	18.45	18.45	18.45	18.45	18.45	18.45	18.45	18.45	18.45	18.45	18.45
4	12.18	12.52	12.67	12.73	12.75	12.75	12.75	12.75	12.75	12.75	12.75	12.75	12.75	12.75	12.75	12.75	12.75	12.75	12.75
5	9.714	10.05	10.24	10.35	10.42	10.46	10.48	10.49	10.49	10.49	10.49	10.49	10.49	10.49	10.49	10.49	10.49	10.49	10.49
6	8.427	8.743	8.932	9.055	9.139	9.198	9.241	9.272	9.294	9.309	9.319	9.325	9.328	9.329	9.329	9.329	9.329	9.329	9.329
7	7.648	7.943	8.127	8.252	8.342	8.409	8.460	8.500	8.530	8.555	8.574	8.589	8.600	8.609	8.616	8.621	8.624	8.626	8.627
8	7.130	7.407	7.584	7.708	7.799	7.869	7.924	7.968	8.004	8.033	8.057	8.078	8.094	8.108	8.119	8.129	8.137	8.143	8.149
9	6.762	7.024	7.195	7.316	7.407	7.478	7.535	7.582	7.619	7.652	7.679	7.702	7.722	7.739	7.753	7.766	7.777	7.786	7.794
10	6.487	6.738	6.902	7.021	7.111	7.182	7.240	7.287	7.327	7.361	7.390	7.415	7.437	7.456	7.472	7.487	7.500	7.511	7.522
11	6.275	6.516	6.676	6.791	6.880	6.950	7.008	7.056	7.097	7.132	7.162	7.188	7.211	7.231	7.250	7.266	7.280	7.293	7.304
12	6.106	6.340	6.494	6.607	6.695	6.765	6.822	6.870	6.911	6.947	6.978	7.005	7.029	7.050	7.069	7.086	7.102	7.116	7.128
13	5.970	6.195	6.346	6.457	6.543	6.612	6.670	6.718	6.759	6.795	6.826	6.854	6.878	6.900	6.920	6.937	6.954	6.968	6.982
14	5.856	6.075	6.223	6.332	6.416	6.485	6.542	6.590	6.631	6.667	6.699	6.727	6.752	6.774	6.794	6.812	6.829	6.844	6.858
15	5.760	5.974	6.119	6.225	6.309	6.377	6.433	6.481	6.522	6.558	6.590	6.619	6.644	6.666	6.687	6.706	6.723	6.739	6.753
16	5.678	5.888	6.030	6.135	6.217	6.284	6.340	6.388	6.429	6.465	6.497	6.525	6.551	6.574	6.595	6.614	6.631	6.647	6.661
17	5.608	5.813	5.953	6.056	6.138	6.204	6.260	6.307	6.348	6.384	6.416	6.444	6.470	6.493	6.514	6.533	6.551	6.567	6.582
18	5.546	5.748	5.886	5.988	6.068	6.134	6.189	6.236	6.277	6.313	6.345	6.373	6.399	6.422	6.443	6.462	6.480	6.497	6.512
19	5.492	5.691	5.826	5.927	6.007	6.072	6.127	6.174	6.214	6.250	6.281	6.310	6.336	6.359	6.380	6.400	6.418	6.434	6.450
20	5.444	5.640	5.774	5.873	5.952	6.017	6.071	6.117	6.158	6.193	6.225	6.254	6.279	6.303	6.324	6.344	6.362	6.379	6.394
24	5.297	5.484	5.612	5.708	5.784	5.846	5.899	5.945	5.984	6.020	6.051	6.079	6.105	6.129	6.150	6.170	6.188	6.205	6.221
30	5.156	5.335	5.457	5.549	5.622	5.682	5.734	5.778	5.817	5.851	5.882	5.910	5.935	5.958	5.980	6.000	6.018	6.036	6.051
40	5.022	5.191	5.308	5.396	5.466	5.524	5.574	5.617	5.654	5.688	5.718	5.745	5.770	5.793	5.814	5.834	5.852	5.869	5.885
60	4.894	5.055	5.166	5.249	5.317	5.372	5.420	5.461	5.498	5.530	5.559	5.586	5.610	5.632	5.653	5.672	5.690	5.707	5.723
120	4.771	4.924	5.029	5.109	5.173	5.226	5.271	5.311	5.346	5.377	5.405	5.431	5.454	5.476	5.496	5.515	5.532	5.549	5.565
∞	4.654	4.798	4.898	4.974	5.034	5.085	5.128	5.166	5.199	5.229	5.256	5.280	5.303	5.324	5.343	5.361	5.378	5.394	5.409

Table 14. Critical Values For Dunnett's Procedure

This table contains critical values $d_{\alpha/2,k,\nu}$ and $d_{\alpha,k,\nu}$ for simultaneous comparisons of each treamtment group with a control group; α is the signficance level, k is the number of treatment groups, and ν is the degrees of freedom of the independent estimate s^2.

Values of $d_{\alpha/2,k,\nu}$ for two-sided comparisons

$\alpha = .05$

ν	k								
	1	2	3	4	5	6	7	8	9
5	2.57	3.03	3.39	3.66	3.88	4.06	4.22	4.36	4.49
6	2.45	2.86	3.18	3.41	3.60	3.75	3.88	4.00	4.11
7	2.36	2.75	3.04	3.24	3.41	3.54	3.66	3.76	3.86
8	2.31	2.67	2.94	3.13	3.28	3.40	3.51	3.60	3.68
9	2.26	2.61	2.86	3.04	3.18	3.29	3.39	3.48	3.55
10	2.23	2.57	2.81	2.97	3.11	3.21	3.31	3.39	3.46
11	2.20	2.53	2.76	2.92	3.05	3.15	3.24	3.31	3.38
12	2.18	2.50	2.72	2.88	3.00	3.10	3.18	3.25	3.32
13	2.16	2.48	2.69	2.84	2.96	3.06	3.14	3.21	3.27
14	2.14	2.46	2.67	2.81	2.93	3.02	3.10	3.17	3.23
15	2.13	2.44	2.64	2.79	2.90	2.99	3.07	3.13	3.19
16	2.12	2.42	2.63	2.77	2.88	2.96	3.04	3.10	3.16
17	2.11	2.41	2.61	2.75	2.85	2.94	3.01	3.08	3.13
18	2.10	2.40	2.59	2.73	2.84	2.92	2.99	3.05	3.11
19	2.09	2.39	2.58	2.72	2.82	2.90	2.97	3.04	3.09
20	2.09	2.38	2.57	2.70	2.81	2.89	2.96	3.02	3.07
24	2.06	2.35	2.53	2.66	2.76	2.84	2.91	2.96	3.01
30	2.04	2.32	2.50	2.62	2.72	2.79	2.86	2.91	2.96
40	2.02	2.29	2.47	2.58	2.67	2.75	2.81	2.86	2.90
60	2.00	2.27	2.43	2.55	2.63	2.70	2.76	2.81	2.85
120	1.98	2.24	2.40	2.51	2.59	2.66	2.71	2.76	2.80
∞	1.96	2.21	2.37	2.47	2.55	2.62	2.67	2.71	2.75

$\alpha = .01$

ν	k								
	1	2	3	4	5	6	7	8	9
5	4.03	4.63	5.09	5.44	5.73	5.97	6.18	6.36	6.53
6	3.71	4.22	4.60	4.88	5.11	5.30	5.47	5.61	5.74
7	3.50	3.95	4.28	4.52	4.71	4.87	5.01	5.13	5.24
8	3.36	3.77	4.06	4.27	4.44	4.58	4.70	4.81	4.90
9	3.25	3.63	3.90	4.09	4.24	4.37	4.48	4.57	4.65
10	3.17	3.53	3.78	3.95	4.10	4.21	4.31	4.40	4.47
11	3.11	3.45	3.68	3.85	3.98	4.09	4.18	4.26	4.33
12	3.05	3.39	3.61	3.76	3.89	3.99	4.08	4.15	4.22
13	3.01	3.33	3.54	3.69	3.81	3.91	3.99	4.06	4.13
14	2.98	3.29	3.49	3.64	3.75	3.84	3.92	3.99	4.05
15	2.95	3.25	3.45	3.59	3.70	3.79	3.86	3.93	3.99
16	2.92	3.22	3.41	3.55	3.65	3.74	3.82	3.88	3.93
17	2.90	3.19	3.38	3.51	3.62	3.70	3.77	3.83	3.89
18	2.88	3.17	3.35	3.48	3.58	3.67	3.74	3.80	3.85
19	2.86	3.15	3.33	3.46	3.55	3.64	3.70	3.76	3.81
20	2.85	3.13	3.31	3.43	3.53	3.61	3.67	3.73	3.78
24	2.80	3.07	3.24	3.36	3.45	3.52	3.58	3.64	3.69
30	2.75	3.01	3.17	3.28	3.37	3.44	3.50	3.55	3.59
40	2.70	2.95	3.10	3.21	3.29	3.36	3.41	3.46	3.50
60	2.66	2.90	3.04	3.14	3.22	3.28	3.33	3.38	3.42
120	2.62	2.84	2.98	3.08	3.15	3.21	3.25	3.30	3.33
∞	2.58	2.79	2.92	3.01	3.08	3.14	3.18	3.22	3.25

Table 14. Critical Values For Dunnett's Procedure (Continued)

Values of $d_{\alpha,k,\nu}$ for one-sided comparisons

$\alpha = .05$

					k				
ν	1	2	3	4	5	6	7	8	9
5	2.02	2.44	2.68	2.85	2.98	3.08	3.16	3.24	3.30
6	1.94	2.34	2.56	2.71	2.83	2.92	3.00	3.07	3.12
7	1.89	2.27	2.48	2.62	2.73	2.82	2.89	2.95	3.01
8	1.86	2.22	2.42	2.55	2.66	2.74	2.81	2.87	2.92
9	1.83	2.18	2.37	2.50	2.60	2.68	2.75	2.81	2.86
10	1.81	2.15	2.34	2.47	2.56	2.64	2.70	2.76	2.81
11	1.80	2.13	2.31	2.44	2.53	2.60	2.67	2.72	2.77
12	1.78	2.11	2.29	2.41	2.50	2.58	2.64	2.69	2.74
13	1.77	2.09	2.27	2.39	2.48	2.55	2.61	2.66	2.71
14	1.76	2.08	2.25	2.37	2.46	2.53	2.59	2.64	2.69
15	1.75	2.07	2.24	2.36	2.44	2.51	2.57	2.62	2.67
16	1.75	2.06	2.23	2.34	2.43	2.50	2.56	2.61	2.65
17	1.74	2.05	2.22	2.33	2.42	2.49	2.54	2.59	2.64
18	1.73	2.04	2.21	2.32	2.41	2.48	2.53	2.58	2.62
19	1.73	2.03	2.20	2.31	2.40	2.47	2.52	2.57	2.61
20	1.72	2.03	2.19	2.30	2.39	2.46	2.51	2.56	2.60
24	1.71	2.01	2.17	2.28	2.36	2.43	2.48	2.53	2.57
30	1.70	1.99	2.15	2.25	2.33	2.40	2.45	2.50	2.54
40	1.68	1.97	2.13	2.23	2.31	2.37	2.42	2.47	2.51
60	1.67	1.95	2.10	2.21	2.28	2.35	2.39	2.44	2.48
120	1.66	1.93	2.08	2.18	2.26	2.32	2.37	2.41	2.45
∞	1.64	1.92	2.06	2.16	2.23	2.29	2.34	2.38	2.42

$\alpha = .01$

					k				
ν	1	2	3	4	5	6	7	8	9
5	3.37	3.90	4.21	4.43	4.60	4.73	4.85	4.94	5.03
6	3.14	3.61	3.88	4.07	4.21	4.33	4.43	4.51	4.59
7	3.00	3.42	3.66	3.83	3.96	4.07	4.15	4.23	4.30
8	2.90	3.29	3.51	3.67	3.79	3.88	3.96	4.03	4.09
9	2.82	3.19	3.40	3.55	3.66	3.75	3.82	3.89	3.94
10	2.76	3.11	3.31	3.45	3.56	3.64	3.71	3.78	3.83
11	2.72	3.06	3.25	3.38	3.48	3.56	3.63	3.69	3.74
12	2.68	3.01	3.19	3.32	3.42	3.50	3.56	3.62	3.67
13	2.65	2.97	3.15	3.27	3.37	3.44	3.51	3.56	3.61
14	2.62	2.94	3.11	3.23	3.32	3.40	3.46	3.51	3.56
15	2.60	2.91	3.08	3.20	3.29	3.36	3.42	3.47	3.52
16	2.58	2.88	3.05	3.17	3.26	3.33	3.39	3.44	3.48
17	2.57	2.86	3.03	3.14	3.23	3.30	3.36	3.41	3.45
18	2.55	2.84	3.01	3.12	3.21	3.27	3.33	3.38	3.42
19	2.54	2.83	2.99	3.10	3.18	3.25	3.31	3.36	3.40
20	2.53	2.81	2.97	3.08	3.17	3.23	3.29	3.34	3.38
24	2.49	2.77	2.92	3.03	3.11	3.17	3.22	3.27	3.31
30	2.46	2.72	2.87	2.97	3.05	3.11	3.16	3.21	3.24
40	2.42	2.68	2.82	2.92	2.99	3.05	3.10	3.14	3.18
60	2.39	2.64	2.78	2.87	2.94	3.00	3.04	3.08	3.12
120	2.36	2.60	2.73	2.82	2.89	2.94	2.99	3.03	3.06
∞	2.33	2.56	2.68	2.77	2.84	2.89	2.93	2.97	3.00

Table 15. Critical Values For Bartlett's Test

This table contains critical values, $b_{\alpha,k,n}$, for Bartlett's test where α is the significance level, k is the number of populations, and n is the sample size from each population.

$\alpha = .05$ k

n	2	3	4	5	6	7	8	9	10
3	.3123	.3058	.3173	.3299	*	*	*	*	*
4	.4780	.4699	.4803	.4921	.5028	.5122	.5204	.5277	.5341
5	.5845	.5762	.5850	.5952	.6045	.6126	.6197	.6260	.6315
6	.6563	.6483	.6559	.6646	.6727	.6798	.6860	.6914	.6961
7	.7075	.7000	.7065	.7142	.7213	.7275	.7329	.7376	.7418
8	.7456	.7387	.7444	.7512	.7574	.7629	.7677	.7719	.7757
9	.7751	.7686	.7737	.7798	.7854	.7903	.7946	.7984	.8017
10	.7984	.7924	.7970	.8025	.8076	.8121	.8160	.8194	.8224
11	.8175	.8118	.8160	.8210	.8257	.8298	.8333	.8365	.8392
12	.8332	.8280	.8317	.8364	.8407	.8444	.8477	.8506	.8531
13	.8465	.8415	.8450	.8493	.8533	.8568	.8598	.8625	.8648
14	.8578	.8532	.8564	.8604	.8641	.8673	.8701	.8726	.8748
15	.8676	.8632	.8662	.8699	.8734	.8764	.8790	.8814	.8834
16	.8761	.8719	.8747	.8782	.8815	.8843	.8868	.8890	.8909
17	.8836	.8796	.8823	.8856	.8886	.8913	.8936	.8957	.8975
18	.8902	.8865	.8890	.8921	.8949	.8975	.8997	.9016	.9033
19	.8961	.8926	.8949	.8979	.9006	.9030	.9051	.9069	.9086
20	.9015	.8980	.9003	.9031	.9057	.9080	.9100	.9117	.9132
21	.9063	.9030	.9051	.9078	.9103	.9124	.9143	.9160	.9175
22	.9106	.9075	.9095	.9120	.9144	.9165	.9183	.9199	.9213
23	.9146	.9116	.9135	.9159	.9182	.9202	.9219	.9235	.9248
24	.9182	.9153	.9172	.9195	.9217	.9236	.9253	.9267	.9280
25	.9216	.9187	.9205	.9228	.9249	.9267	.9283	.9297	.9309
26	.9246	.9219	.9236	.9258	.9278	.9296	.9311	.9325	.9336
27	.9275	.9249	.9265	.9286	.9305	.9322	.9337	.9350	.9361
28	.9301	.9276	.9292	.9312	.9330	.9347	.9361	.9374	.9385
29	.9326	.9301	.9316	.9336	.9354	.9370	.9383	.9396	.9406
30	.9348	.9325	.9340	.9358	.9376	.9391	.9404	.9416	.9426
40	.9513	.9495	.9506	.9520	.9533	.9545	.9555	.9564	.9572
50	.9612	.9597	.9606	.9617	.9628	.9637	.9645	.9652	.9658
60	.9677	.9665	.9672	.9681	.9690	.9698	.9705	.9710	.9716
80	.9758	.9749	.9754	.9761	.9768	.9774	.9779	.9783	.9787
100	.9807	.9799	.9804	.9809	.9815	.9819	.9823	.9827	.9830

Table 15. Critical Values For Bartlett's Test (Continued)

$\alpha = .01$ k

n	2	3	4	5	6	7	8	9	10
3	.1411	.1672	*	*	*	*	*	*	*
4	.2843	.3165	.3475	.3729	.3937	.4110	*	*	*
5	.3984	.4304	.4607	.4850	.5046	.5207	.5343	.5458	.5558
6	.4850	.5149	.5430	.5653	.5832	.5978	.6100	.6204	.6293
7	.5512	.5787	.6045	.6248	.6410	.6542	.6652	.6744	.6824
8	.6031	.6282	.6518	.6704	.6851	.6970	.7069	.7153	.7225
9	.6445	.6676	.6892	.7062	.7197	.7305	.7395	.7471	.7536
10	.6783	.6996	.7195	.7352	.7475	.7575	.7657	.7726	.7786
11	.7063	.7260	.7445	.7590	.7703	.7795	.7871	.7935	.7990
12	.7299	.7483	.7654	.7789	.7894	.7980	.8050	.8109	.8160
13	.7501	.7672	.7832	.7958	.8056	.8135	.8201	.8256	.8303
14	.7674	.7835	.7985	.8103	.8195	.8269	.8330	.8382	.8426
15	.7825	.7977	.8118	.8229	.8315	.8385	.8443	.8491	.8532
16	.7958	.8101	.8235	.8339	.8421	.8486	.8541	.8586	.8625
17	.8076	.8211	.8338	.8436	.8514	.8576	.8627	.8670	.8707
18	.8181	.8309	.8429	.8523	.8596	.8655	.8704	.8745	.8780
19	.8275	.8397	.8512	.8601	.8670	.8727	.8773	.8811	.8845
20	.8360	.8476	.8586	.8671	.8737	.8791	.8835	.8871	.8903
21	.8437	.8548	.8653	.8734	.8797	.8848	.8890	.8926	.8956
22	.8507	.8614	.8714	.8791	.8852	.8901	.8941	.8975	.9004
23	.8571	.8673	.8769	.8844	.8902	.8949	.8988	.9020	.9047
24	.8630	.8728	.8820	.8892	.8948	.8993	.9030	.9061	.9087
25	.8684	.8779	.8867	.8936	.8990	.9034	.9069	.9099	.9124
26	.8734	.8825	.8911	.8977	.9029	.9071	.9105	.9134	.9158
27	.8781	.8869	.8951	.9015	.9065	.9105	.9138	.9166	.9190
28	.8824	.8909	.8988	.9050	.9099	.9138	.9169	.9196	.9219
29	.8864	.8946	.9023	.9083	.9130	.9167	.9198	.9224	.9246
30	.8902	.8981	.9056	.9114	.9159	.9195	.9225	.9250	.9271
40	.9175	.9235	.9291	.9335	.9370	.9397	.9420	.9439	.9455
50	.9339	.9387	.9433	.9468	.9496	.9518	.9536	.9551	.9564
60	.9449	.9489	.9527	.9557	.9580	.9599	.9614	.9626	.9637
80	.9586	.9617	.9646	.9668	.9685	.9699	.9711	.9720	.9728
100	.9669	.9693	.9716	.9734	.9748	.9759	.9769	.9776	.9783

Table 16. Critical Values For Cochran's Test

This table contains critical values, $g_{\alpha,k,n}$, for Cochran's test where α is the significance level, k is the number of independent estimates of variance, each of which is based on ν degrees of freedom.

$\alpha = .05$ ν

k	1	2	3	4	5	6	7	8	9	10	16	36	144	∞
2	.9985	.9750	.9392	.9057	.8772	.8534	.8332	.8159	.8010	.7880	.7341	.6602	.5813	.5000
3	.9669	.8709	.7977	.7457	.7071	.6771	.6530	.6333	.6167	.6025	.5466	.4748	.4031	.3333
4	.9065	.7679	.6841	.6287	.5895	.5598	.5365	.5175	.5017	.4884	.4366	.3720	.3093	.2500
5	.8412	.6838	.5981	.5441	.5065	.4783	.4564	.4387	.4241	.4118	.3645	.3066	.2513	.2000
6	.7808	.6161	.5321	.4803	.4447	.4184	.3980	.3817	.3682	.3568	.3135	.2612	.2119	.1667
7	.7271	.5612	.4800	.4307	.3974	.3726	.3535	.3384	.3259	.3154	.2756	.2278	.1833	.1429
8	.6798	.5157	.4377	.3910	.3595	.3362	.3185	.3043	.2926	.2829	.2462	.2022	.1616	.1250
9	.6385	.4775	.4027	.3584	.3286	.3067	.2901	.2768	.2659	.2568	.2226	.1820	.1446	.1111
10	.6020	.4450	.3733	.3311	.3029	.2823	.2666	.2541	.2439	.2353	.2032	.1655	.1308	.1000
12	.5410	.3924	.3264	.2880	.2624	.2439	.2299	.2187	.2098	.2020	.1737	.1403	.1100	.0833
15	.4709	.3346	.2758	.2419	.2195	.2034	.1911	.1815	.1736	.1671	.1429	.1144	.0889	.0667
20	.3894	.2705	.2205	.1921	.1735	.1602	.1501	.1422	.1357	.1303	.1108	.0879	.0675	.0500
24	.3434	.2354	.1907	.1656	.1493	.1374	.1286	.1216	.1160	.1113	.0942	.0743	.0567	.0417
30	.2929	.1980	.1593	.1377	.1237	.1137	.1061	.1002	.0958	.0921	.0771	.0604	.0457	.0333
40	.2370	.1576	.1259	.1082	.0968	.0887	.0827	.0780	.0745	.0713	.0595	.0462	.0347	.0250
60	.1737	.1131	.0895	.0765	.0682	.0623	.0583	.0552	.0520	.0497	.0411	.0316	.0234	.0167
120	.0998	.0632	.0495	.0419	.0371	.0337	.0312	.0292	.0279	.0266	.0218	.0165	.0120	.0083
∞	0	0	0	0	0	0	0	0	0	0	0	0	0	0

$\alpha = .01$ ν

k	1	2	3	4	5	6	7	8	9	10	16	36	144	∞
2	.9999	.9950	.9794	.9586	.9373	.9172	.8988	.8823	.8674	.8539	.7949	.7067	.6062	.5000
3	.9933	.9423	.8831	.8335	.7933	.7606	.7335	.7107	.6912	.6743	.6059	.5153	.4230	.3333
4	.9676	.8643	.7814	.7212	.6761	.6410	.6129	.5897	.5702	.5536	.4884	.4057	.3251	.2500
5	.9279	.7885	.6957	.6329	.5875	.5531	.5259	.5037	.4854	.4697	.4094	.3351	.2644	.2000
6	.8828	.7218	.6258	.5635	.5195	.4866	.4608	.4401	.4229	.4084	.3529	.2858	.2229	.1667
7	.8376	.6644	.5685	.5080	.4659	.4347	.4105	.3911	.3751	.3616	.3105	.2494	.1929	.1429
8	.7945	.6152	.5209	.4627	.4226	.3932	.3704	.3522	.3373	.3248	.2779	.2214	.1700	.1250
9	.7544	.5727	.4810	.4251	.3870	.3592	.3378	.3207	.3067	.2950	.2514	.1992	.1521	.1111
10	.7175	.5358	.4469	.3934	.3572	.3308	.3106	.2945	.2813	.2704	.2297	.1811	.1376	.1000
12	.6528	.4751	.3919	.3428	.3099	.2861	.2680	.2535	.2419	.2320	.1961	.1535	.1157	.0833
15	.5747	.4069	.3317	.2882	.2593	.2386	.2228	.2104	.2002	.1918	.1612	.1251	.0934	.0667
20	.4799	.3297	.2654	.2288	.2048	.1877	.1748	.1646	.1567	.1501	.1248	.0960	.0709	.0500
24	.4247	.2871	.2295	.1970	.1759	.1608	.1495	.1406	.1338	.1283	.1060	.0810	.0595	.0417
30	.3632	.2412	.1913	.1635	.1454	.1327	.1232	.1157	.1100	.1054	.0867	.0658	.0480	.0333
40	.2940	.1915	.1508	.1281	.1135	.1033	.0957	.0898	.0853	.0816	.0668	.0503	.0363	.0250
60	.2151	.1371	.1069	.0902	.0796	.0722	.0668	.0625	.0594	.0567	.0461	.0344	.0245	.0167
120	.1225	.0759	.0585	.0489	.0429	.0387	.0357	.0334	.0316	.0302	.0242	.0178	.0125	.0083
∞	0	0	0	0	0	0	0	0	0	0	0	0	0	0

Table 17. Critical Values For The Wilcoxon Signed-Rank Statistic

This table contains critical values and probabilities for the Wilcoxon Signed-Rank Statistic T_+; n is the sample size, c_1 and c_2 are defined by $P(T_+ \leq c_1) = \alpha$ and $P(T_+ \geq c_2) = \alpha$.

n	c_1	c_2	α
1	0	1	.500
2	0	3	.250
3	0	6	.125
4	0	10	.062
	1	9	.125
5	0	15	.031
	1	14	.062
	2	13	.094
	3	12	.156
6	0	21	.016
	1	20	.031
	2	19	.047
	3	18	.078
	4	17	.109
7	0	28	.008
	1	27	.016
	2	26	.023
	3	25	.039
	4	24	.055
	5	23	.078
	6	22	.109
8	0	36	.004
	1	35	.008
	2	34	.012
	3	33	.020
	4	32	.027
	5	31	.039
	6	30	.055
	7	29	.074
	8	28	.098
	9	27	.125
9	1	44	.004
	2	43	.006
	3	42	.010
	4	41	.014
	5	40	.020
	6	39	.027
	7	38	.037
	8	37	.049
	9	36	.064
	10	35	.082
	11	34	.102

n	c_1	c_2	α
10	3	52	.005
	4	51	.007
	5	50	.010
	6	49	.014
	7	48	.019
	8	47	.024
	9	46	.032
	10	45	.042
	11	44	.053
	12	43	.065
	13	42	.080
	14	41	.097
	15	40	.116
11	5	61	.005
	6	60	.007
	7	59	.009
	8	58	.012
	9	57	.016
	10	56	.021
	11	55	.027
	12	54	.034
	13	53	.042
	14	52	.051
	15	51	.062
	16	50	.074
	17	49	.087
	18	48	.103
12	7	71	.005
	8	70	.006
	9	69	.008
	10	68	.010
	11	67	.013
	12	66	.017
	13	65	.021
	14	64	.026
	15	63	.032
	16	62	.039
	17	61	.046
	18	60	.055
	19	59	.065
	20	58	.076
	21	57	.088
	22	56	.102

n	c_1	c_2	α
13	9	82	.004
	10	81	.005
	11	80	.007
	12	79	.009
	13	78	.011
	14	77	.013
	15	76	.016
	16	75	.020
	17	74	.024
	18	73	.029
	19	72	.034
	20	71	.040
	21	70	.047
	22	69	.055
	23	68	.064
	24	67	.073
	25	66	.084
	26	65	.095
	27	64	.108
14	12	93	.004
	13	92	.005
	14	91	.007
	15	90	.008
	16	89	.010
	17	88	.012
	18	87	.015
	19	86	.018
	20	85	.021
	21	84	.025
	22	83	.029
	23	82	.034
	24	81	.039
	25	80	.045
	26	79	.052
	27	78	.059
	28	77	.068
	29	76	.077
	30	75	.086
	31	74	.097
	32	73	.108

n	c_1	c_2	α
15	15	105	.004
	16	104	.005
	17	103	.006
	18	102	.008
	19	101	.009
	20	100	.011
	21	99	.013
	22	98	.015
	23	97	.018
	24	96	.021
	25	95	.024
	26	94	.028
	27	93	.032
	28	92	.036
	29	91	.042
	30	90	.047
	31	89	.053
	32	88	.060
	33	87	.068
	34	86	.076
	35	85	.084
	36	84	.094
	37	83	.104
16	19	117	.005
	20	116	.005
	21	115	.007
	22	114	.008
	23	113	.009
	24	112	.011
	25	111	.012
	26	110	.014
	27	109	.017
	28	108	.019
	29	107	.022
	30	106	.025
	31	105	.029
	32	104	.033
	33	103	.037
	34	102	.042
	35	101	.047
	36	100	.052
	37	99	.058
	38	98	.065
	39	97	.072
	40	96	.080
	41	95	.088
	42	94	.096
	43	93	.106

n	c_1	c_2	α	n	c_1	c_2	α	n	c_1	c_2	α	n	c_1	c_2	α
17	23	130	.005	18	27	144	.004	19	32	158	.005	20	37	173	.005
	24	129	.005		28	143	.005		33	157	.005		38	172	.005
	25	128	.006		29	142	.006		34	156	.006		39	171	.006
	26	127	.007		30	141	.007		35	155	.007		40	170	.007
	27	126	.009		31	140	.008		36	154	.008		41	169	.008
	28	125	.010		32	139	.009		37	153	.009		42	168	.009
	29	124	.012		33	138	.010		38	152	.010		43	167	.010
	30	123	.013		34	137	.012		39	151	.011		44	166	.011
	31	122	.015		35	136	.013		40	150	.013		45	165	.012
	32	121	.017		36	135	.015		41	149	.014		46	164	.013
	33	120	.020		37	134	.017		42	148	.016		47	163	.015
	34	119	.022		38	133	.019		43	147	.018		48	162	.016
	35	118	.025		39	132	.022		44	146	.020		49	161	.018
	36	117	.028		40	131	.024		45	145	.022		50	160	.020
	37	116	.032		41	130	.027		46	144	.025		51	159	.022
	38	115	.036		42	129	.030		47	143	.027		52	158	.024
	39	114	.040		43	128	.033		48	142	.030		53	157	.027
	40	113	.044		44	127	.037		49	141	.033		54	156	.029
	41	112	.049		45	126	.041		50	140	.036		55	155	.032
	42	111	.054		46	125	.045		51	139	.040		56	154	.035
	43	110	.060		47	124	.049		52	138	.044		57	153	.038
	44	109	.066		48	123	.054		53	137	.048		58	152	.041
	45	108	.073		49	122	.059		54	136	.052		59	151	.045
	46	107	.080		50	121	.065		55	135	.057		60	150	.049
	47	106	.087		51	120	.071		56	134	.062		61	149	.053
	48	105	.095		52	119	.077		57	133	.067		62	148	.057
	49	104	.103		53	118	.084		58	132	.072		63	147	.062
					54	117	.091		59	131	.078		64	146	.066
					55	116	.098		60	130	.084		65	145	.071
					56	115	.106		61	129	.091		66	144	.077
									62	128	.098		67	143	.082
									63	127	.105		68	142	.088
													69	141	.095
													70	140	.101

Table 18. Critical Values For The Wilcoxon Rank-Sum Statistic

This table contains critical values and probabilities for the Wilcoxon Rank-Sum Statistic W = the sum of the ranks of the m observations in the smaller sample; m and n are the sample sizes, c_1 and c_2 are defined by $P(W \leq c_1) = \alpha$ and $P(W \geq c_2) = \alpha$.

m	n	c_1	c_2	α
2	4	3	11	.067
2	5	3	13	.047
		4	12	.095
2	6	3	15	.036
		4	14	.071
		5	13	.143
2	7	3	17	.028
		4	16	.056
		5	15	.111
2	8	3	19	.022
		4	18	.044
		5	17	.089
		6	16	.133
2	9	3	21	.018
		4	20	.036
		5	19	.073
		6	18	.109
2	10	3	23	.015
		4	22	.030
		5	21	.061
		6	20	.091
		7	19	.136
3	3	6	15	.050
		7	14	.100
3	4	6	18	.028
		7	17	.057
		8	16	.114
3	5	6	21	.018
		7	20	.036
		8	19	.071
		9	18	.125
3	6	6	24	.012
		7	23	.024
		8	22	.048
		9	21	.083
		10	20	.131
3	7	6	27	.008
		7	26	.017
		8	25	.033
		9	24	.058
		10	23	.092
		11	22	.133

m	n	c_1	c_2	α
3	8	6	30	.006
		7	29	.012
		8	28	.024
		9	27	.042
		10	26	.067
		11	25	.097
		12	24	.139
3	9	6	33	.005
		7	32	.009
		8	31	.018
		9	30	.032
		10	29	.050
		11	28	.073
		12	27	.105
3	10	6	36	.003
		7	35	.007
		8	34	.014
		9	33	.024
		10	32	.038
		11	31	.056
		12	30	.080
		13	29	.108
4	4	10	26	.014
		11	25	.029
		12	24	.057
		13	23	.100
4	5	10	30	.008
		11	29	.016
		12	28	.032
		13	27	.056
		14	26	.095
		15	25	.143
4	6	10	34	.005
		11	33	.010
		12	32	.019
		13	31	.033
		14	30	.057
		15	29	.086
		16	28	.129

m	n	c_1	c_2	α
4	7	10	38	.003
		11	37	.006
		12	36	.012
		13	35	.021
		14	34	.036
		15	33	.055
		16	32	.082
		17	31	.115
4	8	10	42	.002
		11	41	.004
		12	40	.008
		13	39	.014
		14	38	.024
		15	37	.036
		16	36	.055
		17	35	.077
		18	34	.107
4	9	10	46	.001
		11	45	.003
		12	44	.006
		13	43	.010
		14	42	.017
		15	41	.025
		16	40	.038
		17	39	.053
		18	38	.074
		19	37	.099
		20	36	.130
4	10	10	50	.001
		11	49	.002
		12	48	.004
		13	47	.007
		14	46	.012
		15	45	.018
		16	44	.026
		17	43	.038
		18	42	.053
		19	41	.071
		20	40	.094
		21	39	.120

m	n	c_1	c_2	α	m	n	c_1	c_2	α	m	n	c_1	c_2	α
5	5	15	40	.004	5	9	16	59	.001	6	7	21	63	.001
		16	39	.008			17	58	.002			22	62	.001
		17	38	.016			18	57	.003			23	61	.002
		18	37	.028			19	56	.006			24	60	.004
		19	36	.048			20	55	.009			25	59	.007
		20	35	.075			21	54	.014			26	58	.011
		21	34	.111			22	53	.021			27	57	.017
5	6	15	45	.002			23	52	.030			28	56	.026
		16	44	.004			24	51	.041			29	55	.037
		17	43	.009			25	50	.056			30	54	.051
		18	42	.015			26	49	.073			31	53	.069
		19	41	.026			27	48	.095			32	52	.090
		20	40	.041			28	47	.120			33	51	.117
		21	39	.063	5	10	16	64	.001	6	8	22	68	.001
		22	38	.089			17	63	.001			23	67	.001
		23	37	.123			18	62	.002			24	66	.002
5	7	15	50	.001			19	61	.004			25	65	.004
		16	49	.003			20	60	.006			26	64	.006
		17	48	.005			21	59	.010			27	63	.010
		18	47	.009			22	58	.014			28	62	.015
		19	46	.015			23	57	.020			29	61	.021
		20	45	.024			24	56	.028			30	60	.030
		21	44	.037			25	55	.038			31	59	.041
		22	43	.053			26	54	.050			32	58	.054
		23	42	.074			27	53	.065			33	57	.071
		24	41	.101			28	52	.082			34	56	.091
5	8	15	55	.001			29	51	.103			35	55	.114
		16	54	.002	6	6	21	57	.001	6	9	23	73	.001
		17	53	.003			22	56	.002			24	72	.001
		18	52	.005			23	55	.004			25	71	.002
		19	51	.009			24	54	.008			26	70	.004
		20	50	.015			25	53	.013			27	69	.006
		21	49	.023			26	52	.021			28	68	.009
		22	48	.033			27	51	.032			29	67	.013
		23	47	.047			28	50	.047			30	66	.018
		24	46	.064			29	49	.066			31	65	.025
		25	45	.085			30	48	.090			32	64	.033
		26	44	.111			31	47	.120			33	63	.044
												34	62	.057
												35	61	.072
												36	60	.091
												37	59	.112

Table 18. Critical Values For The Wilcoxon Rank-Sum Statistic (Continued)

m	n	c_1	c_2	α	m	n	c_1	c_2	α	m	n	c_1	c_2	α
6	10	24	78	.001	7	8	30	82	.001	7	10	32	94	.001
		25	77	.001			31	81	.001			33	93	.001
		26	76	.002			32	80	.002			34	92	.001
		27	75	.004			33	79	.003			35	91	.002
		28	74	.005			34	78	.005			36	90	.003
		29	73	.008			35	77	.007			37	89	.005
		30	72	.011			36	76	.010			38	88	.007
		31	71	.016			37	75	.014			39	87	.009
		32	70	.021			38	74	.020			40	86	.012
		33	69	.028			39	73	.027			41	85	.017
		34	68	.036			40	72	.036			42	84	.022
		35	67	.047			41	71	.047			43	83	.028
		36	66	.059			42	70	.060			44	82	.035
		37	65	.074			43	69	.076			45	81	.044
		38	64	.090			44	68	.095			46	80	.054
		39	63	.110			45	67	.116			47	79	.067
7	7	29	76	.001	7	9	31	88	.001			48	78	.081
		30	75	.001			32	87	.001			49	77	.097
		31	74	.002			33	86	.002			50	76	.115
		32	73	.003			34	85	.003	8	8	39	97	.001
		33	72	.006			35	84	.004			40	96	.001
		34	71	.009			36	83	.006			41	95	.001
		35	70	.013			37	82	.008			42	94	.002
		36	69	.019			38	81	.011			43	93	.003
		37	68	.027			39	80	.016			44	92	.005
		38	67	.036			40	79	.021			45	91	.007
		39	66	.049			41	78	.027			46	90	.010
		40	65	.064			42	77	.036			47	89	.014
		41	64	.082			43	76	.045			48	88	.019
		42	63	.104			44	75	.057			49	87	.025
							45	74	.071			50	86	.032
							46	73	.087			51	85	.041
							47	72	.105			52	84	.052
												53	83	.065
												54	82	.080
												55	81	.097
												56	80	.117

m	n	c_1	c_2	α	m	n	c_1	c_2	α	m	n	c_1	c_2	α
8	9	41	103	.001	9	9	51	120	.001	10	10	64	146	.001
		42	102	.001			52	119	.001			65	145	.001
		43	101	.002			53	118	.001			66	144	.001
		44	100	.003			54	117	.002			67	143	.001
		45	99	.004			55	116	.003			68	142	.002
		46	98	.006			56	115	.004			69	141	.003
		47	97	.008			57	114	.005			70	140	.003
		48	96	.010			58	113	.007			71	139	.004
		49	95	.014			59	112	.009			72	138	.006
		50	94	.018			60	111	.012			73	137	.007
		51	93	.023			61	110	.016			74	136	.009
		52	92	.030			62	109	.020			75	135	.012
		53	91	.037			63	108	.025			76	134	.014
		54	90	.046			64	107	.031			77	133	.018
		55	89	.057			65	106	.039			78	132	.022
		56	88	.069			66	105	.047			79	131	.026
		57	87	.084			67	104	.057			80	130	.032
		58	86	.100			68	103	.068			81	129	.038
8	10	42	110	.001			69	102	.081			82	128	.045
		43	109	.001			70	101	.095			83	127	.053
		44	108	.002			71	100	.111			84	126	.062
		45	107	.002	9	10	53	127	.001			85	125	.072
		46	106	.003			54	126	.001			86	124	.083
		47	105	.004			55	125	.001			87	123	.095
		48	104	.006			56	124	.002			88	122	.109
		49	103	.008			57	123	.003					
		50	102	.010			58	122	.004					
		51	101	.013			59	121	.005					
		52	100	.017			60	120	.007					
		53	99	.022			61	119	.009					
		54	98	.027			62	118	.011					
		55	97	.034			63	117	.014					
		56	96	.042			64	116	.017					
		57	95	.051			65	115	.022					
		58	94	.061			66	114	.027					
		59	93	.073			67	113	.033					
		60	92	.086			68	112	.039					
		61	91	.102			69	111	.047					
							70	110	.056					
							71	109	.067					
							72	108	.078					
							73	107	.091					
							74	106	.106					

Table 19. Critical Values For The Runs Test

This table contains critical values and probabilities for the Runs Test for randomness. Let m be the number of objects of the first kind, n be the number of objects of the second kind ($m \leq n$), and V be the number of runs. The values given are $P(V \leq v)$ in a random arrangement.

m	n	2	3	4	5	6	7	8	9
						v			
2	3	.2000	.5000	.9000	1.0000				
2	4	.1333	.4000	.8000	1.0000				
2	5	.0952	.3333	.7143	1.0000				
2	6	.0714	.2857	.6429	1.0000				
2	7	.0556	.2500	.5833	1.0000				
2	8	.0444	.2222	.5333	1.0000				
2	9	.0364	.2000	.4909	1.0000				
2	10	.0303	.1818	.4545	1.0000				
3	3	.1000	.3000	.7000	.9000	1.0000			
3	4	.0571	.2000	.5429	.8000	.9714	1.0000		
3	5	.0357	.1429	.4286	.7143	.9286	1.0000		
3	6	.0238	.1071	.3452	.6429	.8810	1.0000		
3	7	.0167	.0833	.2833	.5833	.8333	1.0000		
3	8	.0121	.0667	.2364	.5333	.7879	1.0000		
3	9	.0091	.0545	.2000	.4909	.7454	1.0000		
3	10	.0070	.0454	.1713	.4545	.7063	1.0000		
4	4	.0286	.1143	.3714	.6286	.8857	.9714	1.0000	
4	5	.0159	.0714	.2619	.5000	.7857	.9286	.9921	1.0000
4	6	.0095	.0476	.1905	.4048	.6905	.8810	.9762	1.0000
4	7	.0061	.0333	.1424	.3333	.6061	.8333	.9545	1.0000
4	8	.0040	.0242	.1091	.2788	.5333	.7879	.9293	1.0000
4	9	.0028	.0182	.0853	.2364	.4713	.7454	.9021	1.0000
4	10	.0020	.0140	.0679	.2028	.4186	.7063	.8741	1.0000

Table 19. Critical Values For The Runs Test (Continued)

v

m	n	2	3	4	5	6	7	8	9	10	11	12	13	14	15	16	17	18	19	20
5	5	.0079	.0397	.1667	.3571	.6429	.8333	.9603	.9921	1.0000										
5	6	.0043	.0238	.1104	.2619	.5216	.7381	.9112	.9762	.9978	1.0000									
5	7	.0025	.0152	.0758	.1970	.4242	.6515	.8535	.9545	.9924	1.0000									
5	8	.0016	.0101	.0536	.1515	.3473	.5758	.7933	.9293	.9837	1.0000									
5	9	.0010	.0070	.0390	.1189	.2867	.5105	.7343	.9021	.9720	1.0000									
5	10	.0007	.0050	.0290	.0949	.2388	.4545	.6783	.8741	.9580	1.0000									
6	6	.0022	.0130	.0671	.1753	.3918	.6082	.8247	.9329	.9870	.9978	1.0000								
6	7	.0012	.0076	.0425	.1212	.2960	.5000	.7331	.8788	.9662	.9924	.9994	1.0000							
6	8	.0007	.0047	.0280	.0862	.2261	.4126	.6457	.8205	.9371	.9837	.9977	1.0000							
6	9	.0004	.0030	.0190	.0629	.1748	.3427	.5664	.7622	.9021	.9720	.9944	1.0000							
6	10	.0003	.0020	.0132	.0470	.1369	.2867	.4965	.7063	.8636	.9580	.9895	1.0000							
7	7	.0006	.0041	.0251	.0775	.2086	.3834	.6166	.7914	.9225	.9749	.9959	.9994	1.0000						
7	8	.0003	.0023	.0154	.0513	.1492	.2960	.5136	.7040	.8671	.9487	.9879	.9977	.9998	1.0000					
7	9	.0002	.0014	.0098	.0350	.1084	.2308	.4266	.6224	.8059	.9161	.9748	.9944	.9993	1.0000					
7	10	.0001	.0009	.0064	.0245	.0800	.1818	.3546	.5490	.7433	.8794	.9571	.9895	.9981	1.0000					
8	8	.0002	.0012	.0089	.0317	.1002	.2144	.4048	.5952	.7855	.8998	.9683	.9911	.9988	.9998	1.0000				
8	9	.0001	.0007	.0053	.0203	.0687	.1573	.3186	.5000	.7016	.8427	.9394	.9797	.9958	.9993	.99996	1.0000			
8	10	.0000	.0004	.0033	.0134	.0479	.1170	.2514	.4194	.6209	.7822	.9031	.9636	.9905	.9981	.99979	1.0000			
9	9	.0000	.0004	.0030	.0122	.0445	.1090	.2380	.3992	.6008	.7620	.8910	.9555	.9878	.9970	.9997	.99996	1.0000		
9	10	.0000	.0002	.0018	.0076	.0294	.0767	.1786	.3186	.5095	.6814	.8342	.9233	.9742	.9924	.9986	.9998	.99999	1.0000	
10	10	.0000	.0001	.0010	.0045	.0185	.0513	.1276	.2422	.4141	.5859	.7578	.8724	.9487	.9815	.9955	.9990	.9999	.99999	1.0000

Table 20. Tolerance Factors For Normal Distributions

This table contains values of k used to compute tolerance intervals of the form $\bar{x} \pm ks$ for a normal distribution with unknown mean μ and unknown standard deviation σ. The tolerance interval contains at least the proportion $1 - \alpha$ of the population with probability γ.

| | $\gamma = .90$ | | | | $\gamma = .95$ | | | | $\gamma = .99$ | | | |
| | $1 - \alpha$ | | | | $1 - \alpha$ | | | | $1 - \alpha$ | | | |
n	.90	.95	.99	.999	.90	.95	.99	.999	.90	.95	.99	.999
2	15.978	18.800	24.167	30.227	32.019	37.674	48.430	60.573	160.193	188.491	242.300	303.054
3	5.847	6.919	8.974	11.309	8.380	9.916	12.861	16.208	18.930	22.401	29.055	36.616
4	4.166	4.943	6.440	8.149	5.369	6.370	8.299	10.502	9.398	11.150	14.527	18.383
5	3.494	4.152	5.423	6.879	4.275	5.079	6.634	8.415	6.612	7.855	10.260	13.015
6	3.131	3.723	4.870	6.188	3.712	4.414	5.775	7.337	5.337	6.345	8.301	10.548
7	2.902	3.452	4.521	5.750	3.369	4.007	5.248	6.676	4.613	5.488	7.187	9.142
8	2.743	3.264	4.278	5.446	3.136	3.732	4.891	6.226	4.147	4.936	6.468	8.234
9	2.626	3.125	4.098	5.220	2.967	3.532	4.631	5.899	3.822	4.550	5.966	7.600
10	2.535	3.018	3.959	5.046	2.839	3.379	4.433	5.649	3.582	4.265	5.594	7.129
11	2.463	2.933	3.849	4.906	2.737	3.259	4.277	5.452	3.397	4.045	5.308	6.766
12	2.404	2.863	3.758	4.792	2.655	3.162	4.150	5.291	3.250	3.870	5.079	6.477
13	2.355	2.805	3.682	4.697	2.587	3.081	4.044	5.158	3.130	3.727	4.893	6.240
14	2.314	2.756	3.618	4.615	2.529	3.012	3.955	5.045	3.029	3.608	4.737	6.043
15	2.278	2.713	3.562	4.545	2.480	2.954	3.878	4.949	2.945	3.507	4.605	5.876
16	2.246	2.676	3.514	4.484	2.437	2.903	3.812	4.865	2.872	3.421	4.492	5.732
17	2.219	2.643	3.471	4.430	2.400	2.858	3.754	4.791	2.808	3.345	4.393	5.607
18	2.194	2.614	3.433	4.382	2.366	2.819	3.702	4.725	2.753	3.279	4.307	5.497
19	2.172	2.588	3.399	4.339	2.337	2.784	3.656	4.667	2.703	3.221	4.230	5.399
20	2.152	2.564	3.368	4.300	2.310	2.752	3.615	4.614	2.659	3.168	4.161	5.312
21	2.135	2.543	3.340	4.264	2.286	2.723	3.577	4.567	2.620	3.121	4.100	5.234
22	2.118	2.524	3.315	4.232	2.264	2.697	3.543	4.523	2.584	3.078	4.044	5.163
23	2.103	2.506	3.292	4.203	2.244	2.673	3.512	4.484	2.551	3.040	3.993	5.098
24	2.089	2.489	3.270	4.176	2.225	2.651	3.483	4.447	2.522	3.004	3.947	5.039
25	2.077	2.474	3.251	4.151	2.208	2.631	3.457	4.413	2.494	2.972	3.904	4.985
26	2.065	2.460	3.232	4.127	2.193	2.612	3.432	4.382	2.469	2.941	3.865	4.935
27	2.054	2.447	3.215	4.106	2.178	2.595	3.409	4.353	2.446	2.914	3.828	4.888
30	2.025	2.413	3.170	4.049	2.140	2.549	3.350	4.278	2.385	2.841	3.733	4.768
35	1.988	2.368	3.112	3.974	2.090	2.490	3.272	4.179	2.306	2.748	3.611	4.611

Table 20. Tolerance Factors For Normal Distributions (Continued)

n	$\gamma = .90$				$\gamma = .95$				$\gamma = .99$			
	$1 - \alpha$				$1 - \alpha$				$1 - \alpha$			
	.90	.95	.99	.999	.90	.95	.99	.999	.90	.95	.99	.999
40	1.959	2.334	3.066	3.917	2.052	2.445	3.213	4.104	2.247	2.677	3.518	4.493
45	1.935	2.306	3.030	3.871	2.021	2.408	3.165	4.042	2.200	2.621	3.444	4.399
50	1.916	2.284	3.001	3.833	1.996	2.379	3.126	3.993	2.162	2.576	3.385	4.323
55	1.901	2.265	2.976	3.801	1.976	2.354	3.094	3.951	2.130	2.538	3.335	4.260
60	1.887	2.248	2.955	3.774	1.958	2.333	3.066	3.916	2.103	2.506	3.293	4.206
65	1.875	2.235	2.937	3.751	1.943	2.315	3.042	3.886	2.080	2.478	3.257	4.160
70	1.865	2.222	2.920	3.730	1.929	2.299	3.021	3.859	2.060	2.454	3.225	4.120
75	1.856	2.211	2.906	3.712	1.917	2.285	3.002	3.835	2.042	2.433	3.197	4.084
80	1.848	2.202	2.894	3.696	1.907	2.272	2.986	3.814	2.026	2.414	3.173	4.053
85	1.841	2.193	2.882	3.682	1.897	2.261	2.971	3.795	2.012	2.397	3.150	4.024
90	1.834	2.185	2.872	3.669	1.889	2.251	2.958	3.778	1.999	2.382	3.130	3.999
95	1.828	2.178	2.863	3.657	1.881	2.241	2.945	3.763	1.987	2.368	3.112	3.976
100	1.822	2.172	2.854	3.646	1.874	2.233	2.934	3.748	1.977	2.355	3.096	3.954
110	1.813	2.160	2.839	3.626	1.861	2.218	2.915	3.723	1.958	2.333	3.066	3.917
120	1.804	2.150	2.826	3.610	1.850	2.205	2.898	3.702	1.942	2.314	3.041	3.885
130	1.797	2.141	2.814	3.595	1.841	2.194	2.883	3.683	1.928	2.298	3.019	3.857
140	1.791	2.134	2.804	3.582	1.833	2.184	2.870	3.666	1.916	2.283	3.000	3.833
150	1.785	2.127	2.795	3.571	1.825	2.175	2.859	3.652	1.905	2.270	2.983	3.811
160	1.780	2.121	2.787	3.561	1.819	2.167	2.848	3.638	1.896	2.259	2.968	3.792
170	1.775	2.116	2.780	3.552	1.813	2.160	2.839	3.627	1.887	2.248	2.955	3.774
180	1.771	2.111	2.774	3.543	1.808	2.154	2.831	3.616	1.879	2.239	2.942	3.759
190	1.767	2.106	2.768	3.536	1.803	2.148	2.823	3.606	1.872	2.230	2.931	3.744
200	1.764	2.102	2.762	3.529	1.798	2.143	2.816	3.597	1.865	2.222	2.921	3.731
250	1.750	2.085	2.740	3.501	1.780	2.121	2.788	3.561	1.839	2.191	2.880	3.678
300	1.740	2.073	2.725	3.481	1.767	2.106	2.767	3.535	1.820	2.169	2.850	3.641
400	1.726	2.057	2.703	3.453	1.749	2.084	2.739	3.499	1.794	2.138	2.809	3.589
500	1.717	2.046	2.689	3.434	1.737	2.070	2.721	3.475	1.777	2.117	2.783	3.555
600	1.710	2.038	2.678	3.421	1.729	2.060	2.707	3.458	1.764	2.102	2.763	3.530
700	1.705	2.032	2.670	3.411	1.722	2.052	2.697	3.445	1.755	2.091	2.748	3.511
800	1.701	2.027	2.663	3.402	1.717	2.046	2.688	3.434	1.747	2.082	2.736	3.495
900	1.697	2.023	2.658	3.396	1.712	2.040	2.682	3.426	1.741	2.075	2.726	3.483
1000	1.695	2.019	2.654	3.390	1.709	2.036	2.676	3.418	1.736	2.068	2.718	3.472
∞	1.645	1.960	2.576	3.291	1.645	1.960	2.576	3.291	1.645	1.960	2.576	3.291

Table 21. Nonparametric Tolerance Limits

For any distribution of measurements, two-sided tolerance limits are given by the smallest and largest observations in a sample of size n, and a one-sided tolerance limit is given by the smallest (largest) observation in a sample of size n. γ is the probability that the interval will cover a proportion $1 - \alpha$ of the population with a random sample of size n.

$1 - \alpha$ For The Interval Between Sample Extremes

n	γ					
	.5	.7	.9	.95	.99	.995
2	.293	.164	.052	.026	.006	.003
4	.615	.492	.321	.249	.141	.111
6	.736	.640	.490	.419	.295	.254
10	.838	.774	.664	.606	.496	.456
20	.918	.883	.820	.784	.712	.683
40	.959	.941	.907	.887	.846	.829
60	.973	.960	.937	.924	.895	.883
80	.980	.970	.953	.943	.920	.911
100	.984	.976	.962	.954	.936	.929
150	.990	.984	.975	.969	.957	.952
200	.992	.988	.981	.977	.968	.964
300	.995	.992	.988	.985	.979	.976
500	.997	.996	.993	.991	.987	.986
700	.998	.997	.995	.994	.991	.990
900	.999	.998	.996	.995	.993	.992
1000	.999	.998	.997	.996	.994	.993

γ For The Interval Between Sample Extremes

n	$1 - \alpha$				
	.5	.7	.9	.95	.99
2	.250	.090	.010	.003	.000
4	.688	.348	.052	.014	.001
6	.891	.580	.114	.033	.001
10	.989	.851	.264	.086	.004
20	1.000	.992	.608	.264	.017
40		1.000	.920	.601	.061
60			.986	.808	.121
80			.998	.914	.191
100			1.000	.963	.264
150				.996	.443
200				1.000	.595
300					.802
500					.960
700					.993
900					.999
1000					1.000

n For The Interval Between Sample Extremes

$1 - \alpha$	γ					
	.5	.7	.9	.95	.99	.995
.995	336	488	777	947	1325	1483
.99	168	244	388	473	662	740
.95	34	49	77	93	130	146
.90	17	24	38	46	64	72
.85	11	16	25	30	42	47
.80	9	12	18	22	31	34
.75	7	10	15	18	24	27
.70	6	8	12	14	20	22
.60	4	6	9	10	14	16
.50	3	5	7	8	11	12

n For The Interval Below (Above) The Largest (Smallest) Sample Value

$1 - \alpha$	γ				
	.5	.7	.9	.95	.99
.995	139	241	598	919	1379
.99	69	120	299	459	688
.95	14	24	59	90	135
.90	7	12	29	44	66
.85	5	8	19	29	43
.80	4	6	14	21	31
.75	3	5	11	17	25
.65	2	4	9	13	20
.60	2	3	6	10	14
.50	1	2	5	7	10

Table 22. Critical Values For Spearman's Rank Correlation Coefficient

This table contains critical values, $r_{\alpha,n}$, for Spearman's Rank Correlation Coefficient, r_S, where n is the number of pairs of observations and $P(r_S \geq r_{\alpha,n}) = \alpha$.

		α		
n	.05	.01	.005	.001
5	.9000			
6	.8286	.9429		
7	.7143	.8929	.9286	
8	.6429	.8333	.8810	.9524
9	.6000	.7833	.8333	.9167
10	.5636	.7455	.7939	.8788
11	.5364	.7091	.7545	.8455
12	.5035	.6783	.7273	.8182
13	.4835	.6484	.7033	.8022
14	.4637	.6220	.6747	.7758
15	.4429	.6036	.6536	.7536
16	.4294	.5824	.6353	.7324
17	.4142	.5662	.6152	.7108
18	.4014	.5501	.5996	.6945
19	.3912	.5351	.5842	.6772
20	.3805	.5203	.5699	.6617
21	.3701	.5078	.5558	.6481
22	.3608	.4963	.5438	.6341
23	.3528	.4862	.5316	.6215
24	.3443	.4757	.5209	.6087
25	.3369	.4662	.5108	.5977
26	.3306	.4564	.5009	.5870
27	.3236	.4481	.4915	.5763
28	.3175	.4401	.4828	.5670
29	.3118	.4325	.4744	.5576
30	.3063	.4251	.4670	.5488
31	.3012	.4181	.4593	.5403
32	.2962	.4117	.4520	.5323
33	.2914	.4054	.4452	.5247
34	.2871	.3992	.4390	.5172
35	.2826	.3936	.4325	.5101
36	.2788	.3879	.4268	.5035
37	.2748	.3826	.4208	.4969
38	.2710	.3776	.4155	.4905
39	.2674	.3729	.4101	.4846
40	.2640	.3681	.4051	.4788

Acknowledgements

The following tables have been used with permission:

Table 3: Leemis, L. M.(1986), "Relationships Among Common Univariate Distributions," *The American Statistician*, **40**, Number 2, 143-146.

Table 12: Harter, H. Leon(1960), "Tables Of Range And Studentized Range," *Annals of Mathematical Statistics*, **31**, 1122-1147.

Table 13: Harter, H. Leon(1960), "Critical Values For Duncan's New Multiple Range Test," *Biometrics*, **16**, 671-685.

Table 14: Dunnett, Charles W.(1955), "A Multiple Comparison Procedure For Comparing Several Treatments With A Control," *Journal of the American Statistical Association*, **50**, 1096-1121.

Table 15: Dyer, Danny D. and Keating, Jerome P.(1980), "On The Determination Of Critical Values For Bartlett's Test," *Journal of the American Statistical Association*, **75**, 313-319.

Table 16: Eisenhart, C., Hastay, M. W., and Wallis, W. A.(1947), *Techniques of Statistical Analysis*, pages 390-391, New York: McGraw-Hill Book Company.

Table 17: Dixon, Wilfred J. and Massey, Frank J.(1969), *Introduction to Statistical Analysis*, 3rd ed., pages 543-544, New York: McGraw-Hill Book Company.

Table 18: Dixon, Wilfred J. and Massey, Frank J.(1969), *Introduction to Statistical Analysis*, 3rd ed., pages 545-549, New York: McGraw-Hill Book Company.

Table 19: Swed, Frieda S. and Eisenhart, C.(1963), "Tables For Testing Randomness Of Grouping In A Sequence Of Alternatives," *Annals of Mathematical Statistics*, **14**, 66-87.

Table 20: Eisenhart, C., Hastay, M. W., and Wallis, W. A.(1947), *Techniques of Statistical Analysis*, Chapter 2, New York: McGraw-Hill Book Company.

References

Bickel, Peter J. and Doksum, Kjell A.(1977), *Mathematical Statistics: Basic Ideas and Selected Topics*, Oakland, California: Holden-Day, Inc.

Canavos, George C.(1984), *Applied Probability and Statistical Methods*, Boston: Little, Brown and Company.

DeGroot, Morris H.(1986), *Probability and Statistics*, Second Edition, Reading, Massachusetts: Addison-Wesley Publishing Company.

Devore, Jay L.(1987), *Probability and Statistics for Engineering and the Sciences*, Second Edition, Monterey, California: Brooks/Cole Publishing Company.

Draper, N. R. and Smith, H.(1981), *Applied Regression Analysis*, Second Edition, New York: John Wiley & Sons, Inc.

Freund, John E. and Walpole, Ronald E.(1987), *Mathematical Statistics*, Fourth Edition, Englewood Cliffs, NJ: Prentice-Hall, Inc.

Hastings, N. A. J. and Peacock, J. B.(1975), *Statistical Distributions*, London: Butterworth & Company.

Hogg, Robert V. and Craig, Allen T.(1978), *Introduction to Mathematical Statistics*, Fourth Edition, New York: Macmillan Publishing Company, Inc.

Johnson, N. L. and Kotz, S.(1970), *Distributions in Statistics*, Volumes I-IV, New York: John Wiley & Sons, Inc.

Kendall, Maurice, and Stuart, Alan(1977), *The Advanced Theory of Statistics*, Fourth Edition, Volumes I and II, New York: Macmillan Publishing Company, Inc.

LaValle, Irving H.(1970), *An Introduction to Probability, Decision, and Inference*, New York: Holt, Rinehart and Winston, Inc.

Ledermann, Walter (Chief Editor)(1980), *Handbook of Applicable Mathematics*, Volume II, New York: John Wiley & Sons, Inc.

Manoukian, E. B.(1986), *Modern Concepts and Theorems of Mathematical Statistics*, New York: Springer-Verlag.

Mendenhall, William, Scheaffer, Richard L., and Wackerly, Dennis D.(1986), *Mathematical Statistics with Applications*, Third Edition, Boston: Duxbury Press.

Neter, John, Wasserman, William, and Kutner, Michael H.(1985), *Applied Linear Statistical Models*, Second Edition, Homewood, Illinois: Richard D. Irwin, Inc.

Olkin, Ingram, Gleser, Leon J., and Derman, Cyrus(1980), *Probability Models and Applications*, New York: Macmillan Publishing Company.

Patel, J. K., Kapadia, C. H., and Owen, D. B.(1976), *Handbook of Statistical Distributions*, New York: Marcel Dekker.

Snedecor, George W. and Cochran, William G.(1980), *Statistical Methods*, Seventh Edition, Ames, Iowa: The Iowa State University Press.

Walpole, Ronald E. and Myers, Raymond H.(1985), *Probability and Statistics for Engineers and Scientists*, Third Edition, New York: Macmillan Publishing Company.